Amy Brodesky, Amy Doherty, and James Stoddard

Digging into Data with TinkerPlots™

Education Development Center, Inc.

Project Editor	Heather Dever
Editorial Assistant	Christa Edwards
Software Developers	Clifford Konold, Craig D. Miller
Reviewers	Jennifer Buttars, Columbia Elementary School, West Jordan, UT; Marta Johnson, Buncombe County Schools, Asheville, NC; Clifford Konold, University of Massachusetts, Amherst, MA
Data Set Researcher	Stacy Shorr
Accuracy Checkers	Dudley Brooks, Leah Nillas, Cynthia Thomas
Production Director	Casey FitzSimons
Editorial Production Manager	Christine Osborne
Production Editor	Donna Crossman
Copyeditor	David Abel
Production Coordinator/Designer	Thomas Brierly
Compositor	Shepherd Incorporated
Cover Designers	Randall Goodall, Jensen Barnes
Cover Photo Credit	Seth Joel/James Woodson/Mel Yates, Getty Images
Printer	Versa Press, Inc.
Managing Development Editor	Joan Lewis
Publisher	Steven Rasmussen

This material is based upon work supported by the National Science Foundation under award number 9818946. Any opinions, findings, and conclusions or recommendations expressed in this publication are those of the authors and do not necessarily reflect the views of the National Science Foundation.

The Cats data sets in Section 1 were adapted from Friel, S. N. and Joyner, J. M. (Eds.) (1997). *Teach-Stat for Teachers.* Palo Alto, CA: Dale Seymour and are used with permission.

***Digging into Data with TinkerPlots* CD-ROM**

Key Curriculum Press guarantees that the *Digging into Data with TinkerPlots* CD-ROM that accompanies this book is free of defects in materials and workmanship. A defective disk will be replaced free of charge if returned within 90 days of the purchase date. After 90 days, there is a $10.00 replacement fee.

Key Curriculum Press
1150 65th Street
Emeryville, CA 94608
510-595-7000
editorial@keypress.com
www.keypress.com

Printed in the United States of America
10 9 8 7 6 5 4 3 2 13 12 11 10 09 08
ISBN: 978-1-55953-885-5

About the Writers and Field-Testers

Amy Brodesky, M.Ed., has created a variety of mathematics curriculum materials and educational software programs during her 19 years at Education Development Center, Inc., a non-profit organization. She co-authored several middle-school mathematics curriculum units, including *Chance Encounters, Looking Behind the Numbers,* and *From the Ground Up*. For the last five years, she has been the project director for a National Science Foundation grant for the creation of professional development materials to help teachers make mathematics more accessible for a range of students, including learners with special needs.

Amy Doherty, M.Ed., teaches mathematics in the Waltham Public Schools in Massachusetts. She has been teaching at the middle school and high school levels for the past 14 years. Amy has worked closely with several school districts, advising and providing professional development as they implement a standards-based mathematics curriculum. As a teacher participant in the *VISOR: Visualizing Statistical Relationships* project she used TinkerPlots to investigate data analysis questions with other teachers and with her students.

James Stoddard, M.Ed., has taught middle-school mathematics for 15 years in both public and private schools. He is currently working in the Brookline, Massachusetts, public schools, where he also serves as a mentor for new mathematics teachers and as a coordinator of town-wide mathematics competitions. During the summer James co-directs the Math Lab enrichment camp in Groton, Massachusetts.

Digging into Data with TinkerPlots was field-tested over four years in several different classrooms. Formative testing was conducted in 2001 and 2002 by James Stoddard, Edith C. Baker School, Brookline, MA, and Amy Doherty, South Middle School, Waltham, MA. The revised lessons were field-tested in 2003 by Susan Janssen, Heath School, Brookline, MA, as well as by James Stoddard and Amy Doherty. The development team observed students using the lessons, analyzed student work, and interviewed the teachers. The final field-test and evaluation was conducted in 2004 in the classroom of Jullie Rains, Edith C. Baker School, Brookline, MA.

Contents

Contents
continued

Contents

continued

CD Contents

TINKERPLOTS FILES

This folder contains all the data files your students will need to complete the activities. Data sets for suggested extensions are also included for each section. The files are locked as read-only on the CD to prevent students from saving changes in the original, provided files. These locks will be preserved when you copy the files to a Macintosh computer. However, if you are using a Windows computer or a server, the locks will not be preserved. You may want to lock the files on the computer yourself, or put them in a read-only space on the server, so that each student can start with a clean file.

LESSON PDFS

In this folder you'll find a PDF of all the student worksheets, with a bookmarked table of contents for each section. You can use this PDF to print any student worksheet in the book. A PDF of optional worksheets and transparency masters is also included. These optional materials do not appear in the book.

GENERAL PDFS

This folder includes materials you can use in more than one section.

- Correlation of lessons to NCTM standards
- Investigating Data lab sheet: a generic worksheet that can be used with the extensions or as alternative to the lesson worksheets
- Two rubrics for writing high-quality conclusions, both worksheet and transparency forms of each rubric
- Templates for writing conclusions, designed to help students who have difficulty writing
- Vocabulary worksheets for students to write definitions and examples
- Suggested questions to ask students during different parts of the lesson
- Table of data sets for suggested extensions to some lessons

Introduction

To become informed citizens of an increasingly complex and data-driven world, students need to understand statistics. Throughout their lives, knowledge of data analysis will help them see behind the numbers in the news, interpret an abundance of information, and make personal and work decisions.

In the elementary years, students begin studying data analysis by collecting data and representing it with tables, pictographs, bar graphs, and line plots. In middle school, students create and interpret new graphical representations, expand their repertoires of analytical tools, and support their conclusions with strong evidence. In grades six to eight, students begin to think more abstractly and to ask more sophisticated questions about themselves and the world around them. Drawing on these interests, *Digging into Data with TinkerPlots* builds students' knowledge of key data-analysis concepts and skills through investigations of real-world data sets. The lessons provide contexts and opportunities for students to analyze data in depth using a variety of statistical measures and graphical representations. TinkerPlots enables students to analyze the data in ways that go far beyond what they could do with tables and graphs on a static textbook page. Exploring data in dynamic ways, students can quickly change a display from a chart to a graph, alter items of data to answer what-if questions, predict outcomes, and test their reasoning. They learn essential concepts and powerful methods for investigating these central data analysis questions:

- What is typical for a group?
- How do two groups compare?
- What is the relationship between two attributes?
- How do I create, interpret, and compare different graphical representations?
- How do I analyze data to investigate specific questions?
- How do I communicate strong, data-based conclusions?

Digging into Data was designed for and field-tested with students in grades six to eight. Some of the lessons at the beginning of Sections 1 and 2 would also work well with students in grade five. Similarly, some of the lessons at the end of Section 3 could help students in higher grades deepen and apply their understanding of data analysis. All the lessons assume that students have some experience with tables, bar graphs, means, medians, and ranges.

HOW *DIGGING INTO DATA* ADDRESSES NCTM *PRINCIPLES AND STANDARDS FOR SCHOOL MATHEMATICS*

Digging into Data covers the Data Analysis Standards for grades six to eight. Students generate their own questions, make conjectures, and carry out investigations. Every lesson in *Digging into Data* helps students connect a data set to one or more graphical representations of it. They compare the different types of graphs and decide which are better tailored to the question they're trying to answer. *Digging into Data* addresses the expectations below. To see how individual lessons address the standards, see the correlation of lessons to NCTM standards and focal points on the CD.

Data Analysis Standards: In grades six to eight all students should:

- Formulate questions, design studies, and collect data about a characteristic shared by two populations or different characteristics within one population.
- Select, create, and use appropriate graphical representations of data, including histograms, box plots, and scatter plots.
- Find, use, and interpret measures of center and spread, including mean and interquartile range.
- Discuss and understand the correspondence between data sets and their graphical representations, especially histograms, stem-and-leaf plots, box plots, and scatter plots.
- Use observations about differences between two or more samples to make conjectures about the populations from which the samples were taken.
- Make conjectures about possible relationships between two characteristics of a sample on the basis of scatter plots of the data and approximate lines of fit.
- Use conjectures to formulate new questions and plan new studies to answer them.

In *Digging into Data* students apply and deepen their understanding of percentages, ratios, and integers—concepts that are central to the middle-school curriculum and the Number and Operations Standard. Students use percentages to make comparisons within and across groups. Using formulas, they create new attributes that are ratios of other attributes. They apply their knowledge of integers to figure out what positive and negative values mean in the context of specific data sets.

Digging into Data with TinkerPlots provides many opportunities to address the NCTM Communication Standard for grades six to eight. Throughout the lessons, students discuss their hypotheses, findings, and strategies and write strong, specific conclusions that use data as evidence. Using a rubric, they examine sample conclusions, identify areas in need of improvement, and make revisions. By talking and writing about their findings, students consolidate and clarify their thinking about specific data sets and about the data-analysis process.

CHOOSING SECTIONS OR LESSONS

The five sections in *Digging into Data* are designed to be used flexibly to meet the needs of middle-school teachers with different district mathematics objectives and grade levels. You can select one or more sections to use as an addition to your curriculum or as replacement units or chapters or specific lessons to integrate into your curriculum.

It is recommended that you teach a whole section because the mathematics content builds within each section (and across the sections). Sections 1–3 focus on making comparisons of two or more groups. You can choose just one or two sections (1 and 2 or 2 and 3). Sections 4 and 5 each focus on different mathematics concepts and can be taught independently. You can use Section 4 or 5 without doing Sections 1–3 first. Here are the main focuses of each section.

Section 1: Describe and interpret distributions; determine what is typical for a group; and learn about the characteristics of high-quality conclusions.

Section 2: Write and use survey questions to collect data; describe attributes and groups by comparing distributions; create, interpret, and compare box plots; and write strong conclusions based on data.

Section 3: Use percentages to compare groups of different sizes; use formulas to create new attributes; and write conclusions based on data.

Section 4: Use histograms to analyze data and deepen understanding of means and medians.

Section 5: Use scatter plots, ratios, and informal lines of fit to analyze the relationship between two attributes.

If you don't have enough time to teach a section, you can select one or more lessons to address specific objectives in your mathematics program.

- Describe, interpret, and compare distributions of data (Lessons 1.4–1.6, 2.2)
- Determine what is typical for a group (Lessons 1.5, 4.2)
- Make comparisons of two or more groups by using a variety of data-analysis methods (Lessons 1.6, 2.7, 3.3)
- Compare two or more attributes for the same group (Lessons 2.3, 3.2)
- Create, interpret, and compare specific graphical representations (Lessons 1.4–1.6 [line plots], 2.5–2.7 [box plots], 4.5–4.6 [histograms], 5.1–5.5 [scatter plots])
- Build understanding of measures of center (Lessons 4.3, 4.4)
- Explore and analyze the relationship between two attributes (Lessons 3.4, 5.1–5.5)
- Write strong conclusions about data (Lessons 1.7, 3.4, 4.7)
- Collect and analyze own data (Lessons 2.1, 2.4, 3.1)
- Strengthen understanding of ratios in a data-analysis context (Lessons 1.8–1.9, 5.5)
- Apply and extend knowledge of percentages (Lessons 3.3, 4.2)

TEACHING THE LESSONS

In most lessons students build their understanding of data-analysis concepts by using TinkerPlots to actively investigate the data sets and by discussing their findings. The few offline lessons provide an opportunity for you to introduce and connect mathematics concepts. The lessons are structured to help students learn different data-analysis methods and to make productive use of TinkerPlots. (If you would prefer less structure, you can use the Investigating Data worksheet on the CD. Give students the activity question and have them use their own methods to analyze the data.) Based on field-test experience, most lessons can be completed in a 50-minute class period.

Introduction: Take 5–10 minutes to pose the central question and have students write their hypotheses. Writing hypotheses engages students in thinking about the question and helps them invest in analyzing the data. To help students focus, have them open the data set only after they have finished writing their hypotheses. You might also have students write their hypotheses the night before, as homework.

Allow more time for introductions that involve whole-class demonstrations of new TinkerPlots features. For these demonstrations, you may want to use larger fonts to make the graphs easier to read. Go to the **Edit** (Win) **TinkerPlots** (Mac) menu and choose **Preferences.** Check the box for **Use Large Fonts.**

Exploration: Students spend 20–30 minutes investigating the data to answer a question or to build new data-analysis skills. If a pair shares a computer, make sure that each student gets a turn controlling the mouse. If students work at individual computers, they can partner with the student sitting next to them. Build in several different times for partners to show each other their plots and talk about their approaches and findings. It may be helpful to bring the class together briefly to clarify a common difficulty, to demonstrate a TinkerPlots feature, or to spark further investigation. Here are some suggestions for the Exploration.

- As students work, circulate and ask questions that test understanding and prompt deeper thinking. See the Suggested Questions on the CD.
- Establish a rule that students ask each other for help with TinkerPlots before asking you. This gives you more time to talk with students about the data-analysis content and their investigations.
- As you circulate, look for students with different approaches to share with the class in the Wrap-Up.
- Some students may make multiple graphs quickly, without stopping to analyze each graph. Ask questions about one of their graphs to help them slow down and focus.
- Set guidelines for how many graphs students can print and how to choose those graphs. Students should label their printouts with the date and lesson name on their paper or in a text box in the file. Have students staple their graphs to the worksheets.

Wrap-Up: Students spend about 10–15 minutes sharing their findings and the strategies they used to analyze the data. Emphasize that students need to provide evidence from the data to support their conclusions. If you have a computer projection system, have a few students display their plots and explain how they support their conclusions. Some lessons ask students to prepare short presentations of their findings. To help students focus on the

Wrap-Up discussion, have them move away from the computers, turn off the monitors or partially close laptops, or put the keyboard and mouse out of reach.

As an alternative to a whole-class discussion, you could have students discuss their work in small groups. Students could also display a graph on their computer with a text box explaining how the graph supports their conclusions, and then walk around the room looking at the different graphs. Have a class discussion about the similarities and differences in the representations.

If time is running short, it's tempting to skip the Wrap-Up, but this discussion helps to bring together the mathematics ideas in the lesson. If you need to postpone the Wrap-Up, give students a few minutes to list their findings or print a graph so that they can discuss them the next day.

Extensions: Some lessons provide extension questions. Additionally, students could investigate their own questions about data sets they've worked with, using the Investigating Data worksheet on the CD. Several extension data sets are also provided for each section. Each of these data sets contains one or more questions students could explore. There is a table of recommended extensions for each section on the CD.

Homework: *Digging into Data* does not provide specific homework assignments, but here are some tasks that you could assign as homework.

- Take home a printout of a graph and write an explanation of what the graph shows.
- Take home a printout of a table of data. (To make the table, select the data cards and drag a new table from the toolshelf into the document.) Create a graph of one or two attributes by hand.
- Write conclusions using a list of findings from the investigation or revise conclusions using suggestions provided by the teacher or classmates.
- Create reference sheets for the different types of graphs by labeling printouts of graphs, such as a box plot and a histogram. Write a list of tips for making and interpreting each type of graph.
- Create a list of definitions of terms with examples. See the Data Analysis Vocabulary worksheets on the CD.
- Complete a worksheet from an offline lesson.

Assessment: While students are exploring the data, you can informally assess their understanding by circulating and asking questions. You may want to start with more concrete questions to see whether students can read the graphs they create, and then move to questions that involve interpreting the data. It's helpful to ask a variety of questions to get of a sense of different kinds of understandings. See the Suggested Questions on the CD. For small assessment tasks you could give in-class versions of the first two homework suggestions above. Each section has writing assignments that could be used as larger assessment tasks.

- Students write conclusions. (Lessons 1.3, 1.7, 1.8, 1.9, 2.3, 2.4, 2.7, 3.2, 3.3, 3.4, 4.2, 5.4, 5.6)
- Students prepare short presentations. (Lessons 2.4, 4.2)
- Students create a public service announcement. (Lesson 4.6)

Writing about Data: *Digging into Data* provides many opportunities for students to write about data and to improve their communication skills. See above for specific examples.

Sample student work from the field test is provided in some lessons to give you an idea of what students might write. The *Digging into Data* CD has a variety of templates to give students additional support for writing conclusions. These are useful for students who feel overwhelmed by the task of filling a blank page and for others who have difficulty with writing. If your school's language arts classes use writing templates, consider using the same ones for writing conclusions in math.

Digging into Data uses rubrics to help you set clear expectations for writing and define the characteristics of high-quality conclusions. Students also benefit from looking at examples of other students' work (see Lesson 1.7). There are two generic rubrics for conclusions on the CD: one for making comparisons and one for analyzing the relationship between attributes. You can use them as is or as a starting point for creating your own rubrics. If students are new to using rubrics, you may want to focus on one or two criteria at a time, adding more criteria in later lessons. Some lessons have specific versions of the rubrics.

USING TINKERPLOTS

Getting Started: There are several ways to learn how to use TinkerPlots. Students can view the short introductory movie, *TinkerPlots Basics,* available from the TinkerPlots **Help** menu. Alternatively, you may want to do a class demonstration of the basic features using a computer projection system. Students tend to learn how to use TinkerPlots quickly. Have a few copies of the Quick Reference cards available for students. (To print additional Quick Reference cards, go to the **TinkerPlots** application folder, then **TinkerPlots Help | Quick Reference.pdf.**) It's helpful to give students time to experiment with different features. They will then be better able to focus on specific tasks in the lessons.

TinkerPlots Skills: The lesson notes list prerequisite and new TinkerPlots skills. The prerequisite skills for graphing are grouped into two categories: basic and intermediate. Most of the lessons assume that students have basic graphing skills.

Basic graphing skills: Students should be able to make and explore graphs by using the **Separate, Order,** and **Stack** buttons, dragging and dropping attributes, undoing, changing icon type and size, and selecting cases. These skills are explained in the *TinkerPlots Basics* movie.

Intermediate graphing skills: Students should be able to add features to their plots by showing measures of center, counts, and percentages, and using dividers, the **Drawing** tool, and reference lines. These skills are explained as needed in the lessons.

Managing Screen Space and Printing: With TinkerPlots, students can choose to replace one graph with another or to display several graphs at once. Some students may have difficulty organizing their screen space both while they are working and for printing. Here are some suggestions.

- Work with just one or two graphs at a time. Delete graphs you don't need.
- Try not to overlap objects. In particular, don't place a text box on top of a graph (a title, for example). When you select the graph, the text box will be hidden.
- *Iconify* graphs or other objects by shrinking them until they become icons.
- To hide a graph (or other object), select the graph and choose **Hide Plot** from the **Edit** menu. Show the graph by choosing **Show Hidden Objects** from the **Edit** menu.
- Before printing, go to the **File** menu and choose **Show Page Breaks** (Win or Mac) or **Print Preview** (Win).

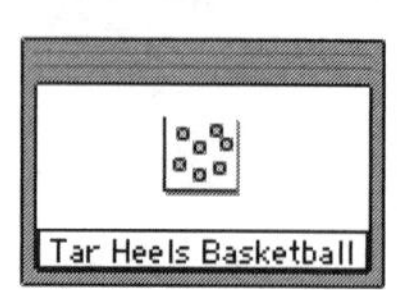

Digging into Data with TinkerPlots™

Section 1

Comparisons, Distributions, and Line Plots: Exploring Data about Cats

In this section, students are introduced to the basic idea of data analysis—using data to answer a question. The section focuses on using different types of graphs, particularly line plots, to compare groups. Students learn many different ways to compare distributions, including looking at shape, finding center clumps, and making numeric comparisons. Students also work with ratios to answer their questions. It is not necessary to use all nine lessons in this section. If your students are new to data anlysis, consider using Lessons 1.1–1.7, omitting the work on ratios in Lessons 1.8 and 1.9. If your students are familiar with line plots, distributions, medians, and ratios, consider using Lessons 1.3 and 1.5–1.9.

OBJECTIVES

Understanding Attributes and Distributions

- Identify which attributes to analyze to investigate a question about an individual case or a group of cases (Lessons 1.2, 1.3, 1.9)
- Recognize the value of analyzing multiple attributes to get a fuller picture of the data (Lessons 1.3, 1.9)
- Describe and interpret the shapes of distributions (Lessons 1.4–1.6, 1.8, 1.9)
- Reason about what the shapes of distributions might be for different attributes (Lessons 1.2, 1.3)

Analyzing Data

- Connect a point on a plot with a value for a case (Lessons 1.2, 1.3, 1.8, 1.9)
- Create and interpret a variety of plots to analyze one or more attributes (Lessons 1.3, 1.6)

- Create, interpret, and compare line plots (Lessons 1.4–1.6, 1.8, 1.9)
- Move from focusing on individual cases to developing an aggregate view of the data (Lessons 1.5, 1.6)
- Identify what's "typical" for a group (Lessons 1.5, 1.6)
- Compare one case to a group of cases (Lessons 1.3, 1.9)
- Compare two groups (Lessons 1.6, 1.9)
- Use statistical measures of center and spread (mean, median, and range) to make comparisons (Lessons 1.4, 1.6)

Communicating about Data

- Describe findings about a group of cases (Lessons 1.4, 1.5)
- Write comparison statements to fairly compare two groups (Lessons 1.7, 1.9)
- Write clear conclusions that use the data as evidence (Lessons 1.7–1.9)
- Build understanding of the characteristics of high-quality conclusions (Lesson 1.7)

Applications of Math Concepts from Other Strands

- Number and Operations: Use percentages to analyze and compare data sets (Lessons 1.4–1.6)
- Number and Operations: Interpret and compare ratios (Lessons 1.8, 1.9)
- Algebra: Create formulas to determine ratios (Lessons 1.8, 1.9)
- Measurement: Deepen and apply knowledge of measures of length (inches) and weight (pounds) (Lessons 1.1, 1.3, 1.8, 1.9)

TINKERPLOTS SKILLS

No previous TinkerPlots experience is needed for this section. In Lesson 1.2, students learn what they need to know to make and explore graphs. In this section, students also learn how to add graph features such as drawings, dividers, measures of center, and percentages. In Lessons 1.8 and 1.9, they create a new attribute and define it with a formula to help them work with the ratios.

OVERVIEW

This activity is designed to introduce students to the data card format and to the attributes they will be investigating. It is important for students to understand how the data were collected and what the attribute names mean before they begin investigating the data.

Objectives

- Become familiar with the term *attribute*
- Become familiar with the context of the cats data
- Become familiar with the TinkerPlots data card format
- Read information from data cards, pose questions, and plan studies

Offline **Class Time:** One class period

Materials

- Cats and Attributes worksheet (one per student)
- Chart paper (*optional*)
- Sticky notes (*optional*)

LESSON PLAN

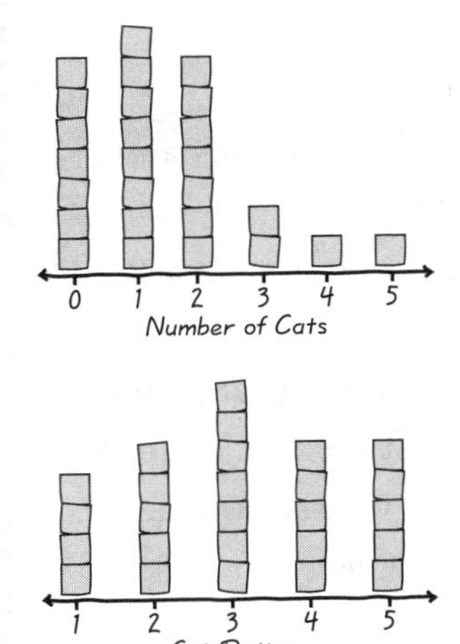

Introduction

1. Start the lesson by collecting some data from the class about cats and other pets. One way to collect and display the data is to have students put their responses on sticky notes and then put their sticky notes on a large line plot on chart paper. Use two or three of these questions so that students can make comparisons across graphs.
   - How many cats do you have?
   - How would you rate cats as pets? (Use a scale of 1 to 5, with 1 = terrible pets; 3 = OK pets; 5 = great pets)
   - How would you rate dogs as pets? (Use a scale of 1 to 5, with 1 = terrible pets; 3 = OK pets; 5 = great pets)

2. Have students examine the graphs and write down three things that they found out about the class from the data. Then have a whole-class discussion by using some of these questions. (If you do this activity with more than one class section, then have students compare their section's results with another section's as well.)
   - In our class, how many cats do students typically have?
   - Is it common for students in our class to have a cat? Explain.
   - How much agreement or disagreement is there in the class's ratings of cats as pets?
   - Overall, does our class think that dogs or cats make better pets?

Exploration

3. Explain that students will be investigating data about different attributes, or characteristics, of cats, such as weight and body length. Introduce the term *attribute.* Have students brainstorm a list of attributes about cats.
4. Hand out copies of the Cats and Attributes worksheet. Explain that the cats data were collected by pet owners—they took measurements of their cats. Introduce the attributes on the data cards in question 1.
5. Have students work individually or in pairs to read the data cards and write their own questions that can be answered by using the data on the two cards (question 1). Have students share some questions and answers with the class.
6. Ask students to make predictions about the sizes of cats in the data set (question 2). The process of making predictions helps students become more invested in finding out the information in the data set. To help students have a sense of the sizes of cats, ask them to find things in the classroom that are about the length of the example cats on the data cards: 18 in. and 14 in. As points of reference, students can think about the length of a piece of paper (11 in.), and of their school desks. Similarly, students can use the weights of different objects, such as a 5-lb bag of flour, to think about the weights of cats.
7. Have students work in pairs or small groups to brainstorm attributes for the topics in questions 3 and 4.

Wrap-Up

8. Have a class discussion about the attributes that students brainstormed for different topics.
   - What topic did you choose? What attributes did you brainstorm for that topic?
9. Point out that in this lesson students have worked on parts of the data analysis process.
   - Ask a Question: They worked with different questions about cats and came up with their own questions about attributes.
   - Collect Data: They collected data by taking a class survey about pets.
   - Analyze Data: They made a graph of the class survey data and analyzed it.
   - Communicate Conclusions: They discussed their findings about the class survey data on pets.
10. Ask students to choose one of the attributes from their chart and think about how they would investigate the data (question 4). This is an opportunity for students to connect to their previous experiences investigating data. Remind students to be respectful of what others might consider to be private information, such as grades. Have students share their ideas with the class.
    - What questions would you like to investigate about the attribute you chose?
    - How would you collect the data for your attribute?
    - What do you think you would find out? (What is your hypothesis?)

ANSWERS

Cats and Attributes

1. Sample answers:

   a. Which cat has the longer tail? Blacky's tail is 2 in. longer.

   b. Are Blacky and Gabriel the same gender? No

   c. Which cat has the longer body? Blacky's body is 4 in. longer.

2. Sample answers:

Attribute	Lowest value	Highest value
Weight (lb)	1	20
Body length (in.)	5	25
Tail length (in.)	4	18

3. Sample student work:

Topic	Attributes with numbers as values	Attributes with words as values
a. Middle-school students	D.J.: Hours of homework D.J.: GPA's Peter: Hours on computer or t.v. per day Peter: Hours exercising per day	D.J.: Favorite subjects D.J.: Favorite sports Eana: Hair color Eana: Favorite clothing
b. D.J.: Movies	Box Office Gross Budgets for movies	Favorite Movies Favorite Actors
b. Peter: Dogs	weight to length ratio age	name country of origin

4. Answers will vary.

Cats and Attributes

Name:

An *attribute* is a characteristic of a person or thing. *Variable* is another term for attribute. In this unit you will use TinkerPlots to investigate data about cats. Pet owners collected data about their cats for seven attributes: name, age, gender, weight, body length, tail length, and eye color. (Body length was measured from the cat's head to the base of its tail. It does not include the tail.)

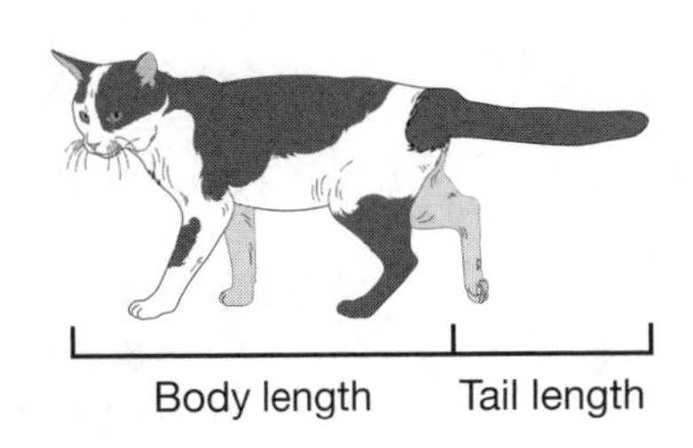

1. In TinkerPlots, the information for each cat is shown on a data card. Here are the data cards for two cats.

One-Year-Old Cats

case 1 of 20

Attribute	Value	Unit
Name	Blacky	
Gender	Female	
Weight	5	pounds
Body_Length	18	inches
Tail_Length	12	inches
Eye_Color	Yellow	

One-Year-Old Cats

case 14 of 20

Attribute	Value	Unit
Name	Gabriel	
Gender	Male	
Weight	7	pounds
Body_Length	14	inches
Tail_Length	10	inches
Eye_Color	Blue	

Write and answer two questions to compare the two cats, Blacky and Gabriel. Use the information on the data cards to answer your questions.

Question	Answer
Which cat weighs more and by how much?	*Gabriel weighs 2 more pounds than Blacky.*
a.	
b.	

2. You will be looking at a data set of 20 one-year-old cats. What do you think the highest and lowest values will be for each attribute?

Attribute	Lowest value	Highest value
Weight (lb)		
Body length (in.)		
Tail length (in.)		

3. What other attributes and topics would you like to collect data about? For part b, add your own topic.

Topic	Attributes with numbers as values	Attributes with words as values
a. Middle-school students	• *Number of pets* • •	• *Favorite music groups* • •
b.	• • •	• • •

4. Make a plan for investigating one of the attributes from your chart. Write your answers on your own paper.

   a. Ask a Question: Which attribute do you want to investigate? What do you want to find out about the attribute?

   b. Collect Data: You need to gather data from 100 people. Who would you ask? How would you collect the data? For example, you could take a survey of eighth-graders.

   c. Analyze Data: What would you do to answer your question?

   d. Communicate Conclusions: Who would be interested in hearing about your findings? What do you think the data will show? Why?

Exploring Graphs and Attributes

OVERVIEW

The Finding Cats activity provides a structured way for students to become familiar with TinkerPlots. Because TinkerPlots is a very open-ended tool, the activity gives students a concrete task to help them get comfortable. The task of finding maximum and minimum values is a good match to students' tendencies to look for extreme cases in a data set. The data set has only 20 cats because this is a manageable number of cases for students to investigate. The number of dots or bars is not overwhelming for students. The activity also helps students to make connections between a dot on a graph and the corresponding data card.

Objectives

- Become familiar with using TinkerPlots
- Determine which attributes to use to answer specific questions
- Display and analyze two attributes at the same time
- Find highest (maximum) and lowest (minimum) values in a data set
- Understand how a particular case is represented in a graph

TinkerPlots

Class Time: One class period. If time is short, you can skip the Name the Mystery Attributes activity. Students will have the opportunity to do a similar activity in Lesson 1.3.

Materials

- Finding Cats worksheet (one per student)
- Name the Mystery Attributes worksheet (one per student)

Data Sets: One Year Old Cats.tp (20 one-year-old cats), **Mystery Attributes 1.tp**

TinkerPlots Skills: Basic graphing is explained in this lesson.

LESSON PLAN

Introduction

1. Introduce TinkerPlots to students. Here are two possible ways.
   - Do a demonstration for the class using a projection system. Introduce the basic features of TinkerPlots by showing how to find the cat with the heaviest weight. Then show students how to display two attributes at the same time to answer the question: Who is the heaviest female cat? Go over these features: separate, order, stack, mix-up, undo, how to select attributes, and how to exchange one attribute for another. You may want to watch the movie *TinkerPlots Basics* to help you plan your demonstration. To watch the movie, open TinkerPlots. Choose **TinkerPlots Movies** from the **Help** menu.
   - Have students watch the *TinkerPlots Basics* movie individually, or as a class on a projection system.
2. Introduce the task of finding cats with particular attributes. Point out that there may be more than one cat for an answer.

Exploration

3. Students use TinkerPlots to find cats with specific attributes. They also come up with their own questions for finding cats (questions 1–8). Emphasize that students should make plots to answer the questions rather than clicking through the data cards. Encourage students to experiment with finding different ways to make plots, so that they can get to know the software.
4. After students have finished, bring the class together to go over the answers. Have students demonstrate how they found the answers to some of the questions that have two attributes. It's helpful to have students show different ways to plot the data to find the answer to the same question. There are many ways to use TinkerPlots to show two attributes at the same time. Here are three examples for question 3: What is the weight of the lightest male cat?
   - Fully separate the data by *Weight.* Then show *Gender* with color.
   - Separate the data by *Weight* and by *Gender.*

- Use value bars to represent the weights. (Choose **Value Bar Vertical** or **Horizontal** from the **Icon Type** menu.) Order the weights from smallest to largest. Then color by *Gender.*

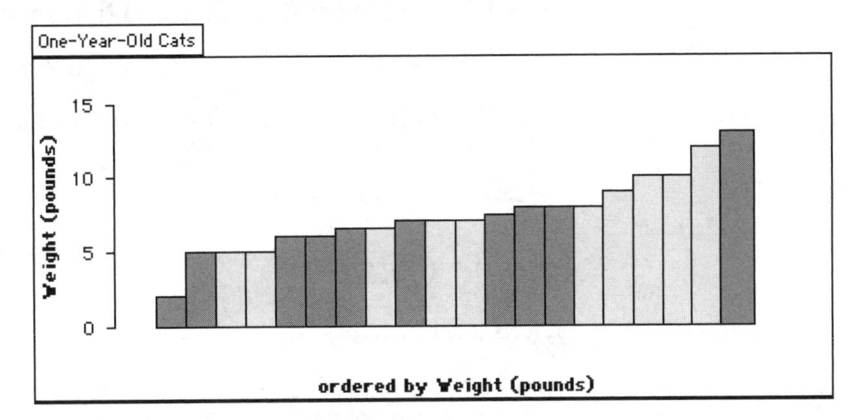

5. Have one or two students pose the questions they wrote, and have the class find the answers. Discussion questions:
   - How did you plot the data to find the cats? Which attribute or attributes did you plot?
   - What other ways could you plot the data to find the cats?
6. Introduce the Name the Mystery Attributes activity. Students should open the file **Mystery Attributes 1.tp.** The file shows four different graphs but does not reveal the names of the attributes. Demonstrate how to scroll to see the different graphs. Students may also need to move the data card: select it and drag the top border. Students need to figure out which attribute each graph represents, and then create the graph with TinkerPlots. This activity helps students learn how to use the software to create a variety of graphs, and builds their familiarity with the attributes in the cats data set. Here are some questions to help students figure out what the mystery attribute might be.
   - What are the highest and lowest values on the graph? Which of the attributes is likely to have this range of values?
   - What does the shape of the data tell you? Which attribute is likely to have that distribution of data?

Wrap-Up

7. Discuss the strategies that students used in the activities.
   - What strategies did you use to find specific cats?
   - What strategies did you use to figure out the mystery attributes?
   - What are some of the things that you learned to do with TinkerPlots?

ANSWERS

Finding Cats

1. 13 in.; Fuzzy
2. 12 in.; Keikko, Blacky
3. 5 lb; Gray, Ashes
4. 14 in.; Tomadachi Joto, Gabriel
5. 11 in.; Chelsea

6., 7. Sample student work:

Emily: How many female cats have blue eyes? 1; Keikko

Justin: What is the body length of the heaviest cat? 18 inches; Chelsea

8. 11 in.; Teak (To show three attributes at the same time you can, for example, put *Weight* on the horizontal axis, put *Tail_Length* on the vertical axis, and color by *Gender*.)

Name the Mystery Attributes

1. *Body_Length*
2. *Weight*
3. *Tail_Length*
4. *Body_Length*

Finding Cats

Name:

In this lesson you'll look for specific cats in the data set. Open the TinkerPlots file **One Year Old Cats.tp** to see data about 20 one-year-old cats. Use TinkerPlots to make plots that help you answer the questions in the table. *Note:* There may be more than one cat for an answer.

Question	Answer	Name of cat(s)
Looking at one attribute		
What is the heaviest weight?	*13 pounds*	*Chelsea*
1. What is the shortest body length?	_______ inches	
2. What is the longest tail length?	_______ inches	
Looking at two attributes		
3. What is the weight of the lightest male cat?	_______ pounds	
4. What is the body length of the shortest male cat?	_______ inches	
5. What is the tail length of the heaviest cat?	_______ inches	
Write your own questions		
6.	_______	
7.	_______	
Challenge: Looking at three attributes		
8. What is the tail length of the heaviest male cat?	_______ inches	

Name the Mystery Attributes

Name:

Open the TinkerPlots file **Mystery Attributes 1.tp.** The graphs in this file show data from the data set of 20 one-year-old cats. (Scroll down to see the different graphs. If you need to move the data card, select it and drag the top border.) For each graph, you need to figure out what attribute(s) the graph shows and then make the graph with TinkerPlots.

1. What is mystery attribute A?

   Name Gender Weight Body_Length Tail_Length Eye_Color

   Make the graph with TinkerPlots.

2. What is mystery attribute B?

   Name Gender Weight Body_Length Tail_Length Eye_Color

   Make the graph with TinkerPlots.

3. Which two attributes does this graph show?

   Gender and __________

   Make the graph with TinkerPlots.

4. Which two attributes does this graph show?

   Gender and _________

   Make the graph with TinkerPlots.

Comparing an Individual to a Group

OVERVIEW

This lesson provides a useful way for you to get a sense of students' initial thinking about making comparisons. Students need to determine whether the cat named Chubbs is chubby in comparison to the group of cats. Many students base their conclusions about Chubbs on one attribute: weight. Usually, a few students examine both weight and body length, and make the case that it is important to consider both of these attributes to determine if a cat is chubby. This brings up one of the essential messages of this section—by analyzing more than one attribute you get a fuller picture of the data and are able to make more comprehensive comparisons.

Objectives

- Compare an individual case to a group
- Understand how a particular case is represented on a graph
- Experiment with creating plots to make comparisons

TinkerPlots

Class Time: One class period. If time is short, you can skip the Mystery Attributes part of the activity.

Materials

- Is the Cat Named Chubbs Chubby? worksheet (one per student)

Data Sets: **One Year Old Cats.tp** (20 one-year-old cats), **Female Cats.tp** (50 female cats of different ages), **Mystery Attributes 2.tp**

TinkerPlots Prerequisites: Student should be familiar with basic graphing.

Assessment Notes: The Chubbs activity is meant to be an informal preassessment of these questions:

- Do students know how to choose the appropriate attribute(s) to make a comparison?
- Do students use one or two attributes to make the comparison?
- Do they recognize the importance of considering both weight and body length?
- What kinds of plots do students create to make the comparison?

To find out students' initial understandings, ask them to work independently and to figure out for themselves how to make the comparisons and plot the data. At the end of this section, students can return to a similar question about a cat named Feather, giving them an opportunity to see what they have learned.

LESSON PLAN

Introduction

1. Introduce the question: Is the cat named Chubbs chubby compared to this group of one-year-old cats? Point out that all the cats in the data set are about the same age as Chubbs. Emphasize that students need to compare Chubbs to the group of cats in the data set (not to compare Chubbs to just one other cat or to cats in general). Their conclusions need to be based on the evidence in this data set.

Exploration

2. The first task is finding Chubbs in the data set (question 2). Encourage students to make plots to find Chubbs rather than browsing through the data cards. If students don't know where to begin, suggest they make a graph of the information given—body length and gender.
3. Students create their own plots and analyze the data (questions 3–6). They draw a sketch of a plot that they found helpful and write their conclusions. As an alternative to drawing the plots, the students could print out their graphs and attach them to the worksheet.
4. Discuss students' conclusions. Have one or two students show their plots to the class.
   - What is your conclusion? Is Chubbs chubby compared to this group of one-year-old cats?
   - What attribute or attributes did you compare?
   - How did you plot the data?
5. Introduce the new data set that students will be using for the nicknames task. This data set has 50 female cats of different ages. Direct students to open the file **Female Cats.tp.**

6. Introduce the nicknames task. In the previous part of the activity, students found that the name of Chubbs was not a good match to the cat. Here, students work independently or in pairs to pick a nickname and find a cat in the data set that they think is a good match. They need to figure out which attributes to plot to find a cat that is a good match to the nickname (question 8). Then they come up with their own nickname for a cat (question 9).

7. If you want students to look at the mystery graphs (questions 10 and 11), have them open the file **Mystery Attributes 2.tp.** Remind them to scroll down to see the different graphs and move the data card as needed.

Wrap-Up

8. Discuss students' strategies for the nicknames task.
   - What strategies did you use to figure out which cat was a good match to each nickname? Which attributes did you look at?
   - What nickname did you come up with for the cat? Why is this nickname a good match to this particular cat?

ANSWERS

Is the Cat Named Chubbs Chubby?

2. Chubbs weighs 7 lb. This is a helpful plot for finding Chubbs. It shows two attributes: *Body_Length* is on the horizontal axis and *Gender* is shown by color. Chubbs is the only male cat with a body length of 22 in.

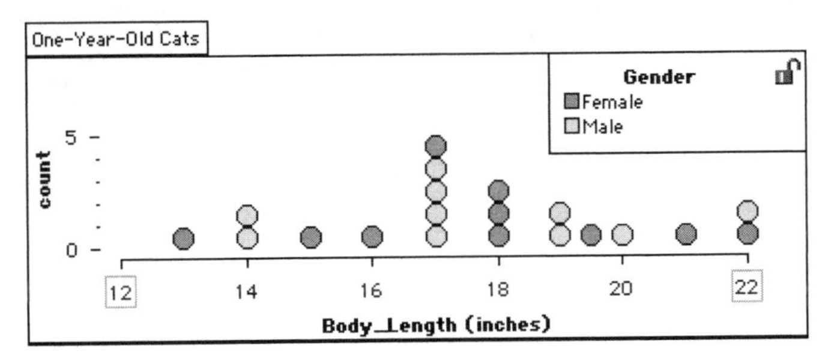

3.–5. Some students horizontally stack the data by weight and order it from lightest to heaviest. This shows that Chubbs is in the middle of the weights.

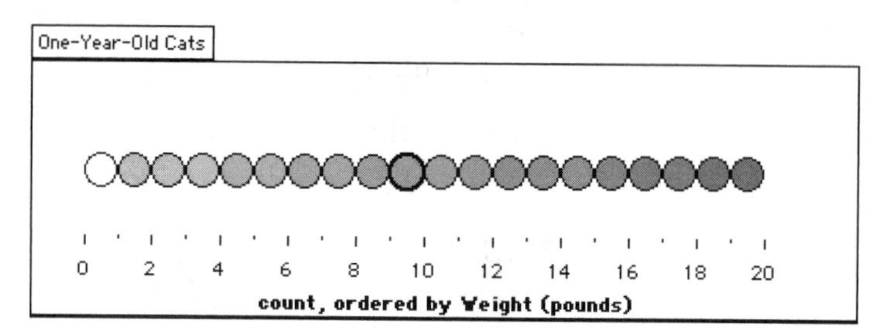

Some students make an additional plot for body length and find that Chubbs is one of the longest cats. The combination of the two plots shows that Chubbs has a medium weight and a long body length, so he is not chubby.

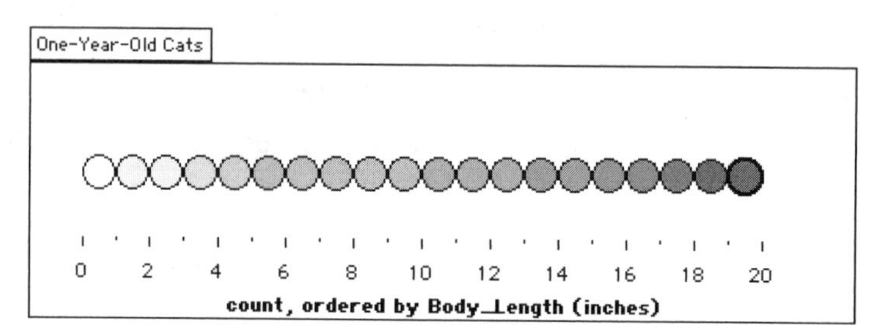

Some students also separate the data by gender to compare Chubbs to the other male cats.

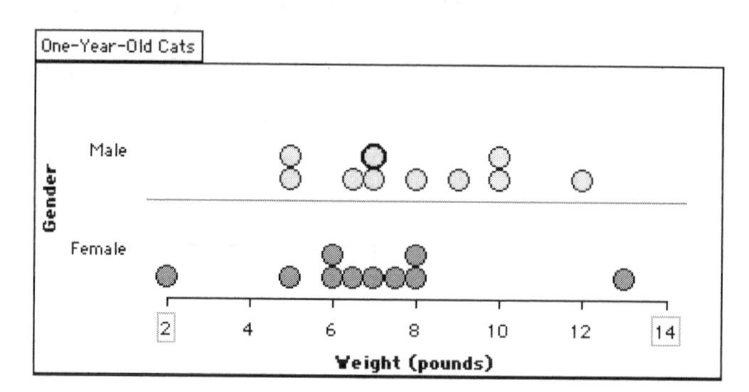

6. Sample student work:

Ian: *Chubbs is not chubby because he is the longest cat, but he is medium weight*

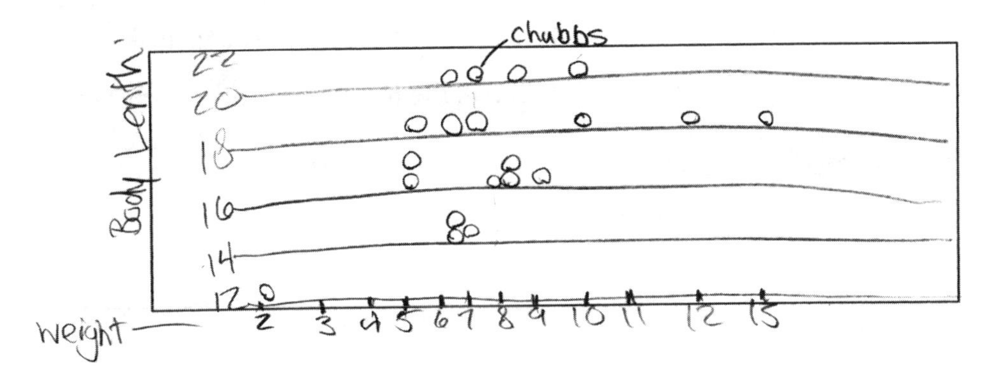

Comments: This student considers two attributes, weight and body length, for making conclusions. He displays both attributes at the same time on his plot, which looks similar to a scatter plot.

Eana: *My conclusion is that Chubbs is not really that chubby because for weight he's not really that heavy compared to the ones that weigh more than he does.*

Comments: The student considers only one attribute: weight. Her plot shows that she stacked all the weights in order and then circled Chubbs. Her conclusion is correct but her explanation is hard to follow.

Alexei: *My conclusion is that Chubbs is not chubby because ?*

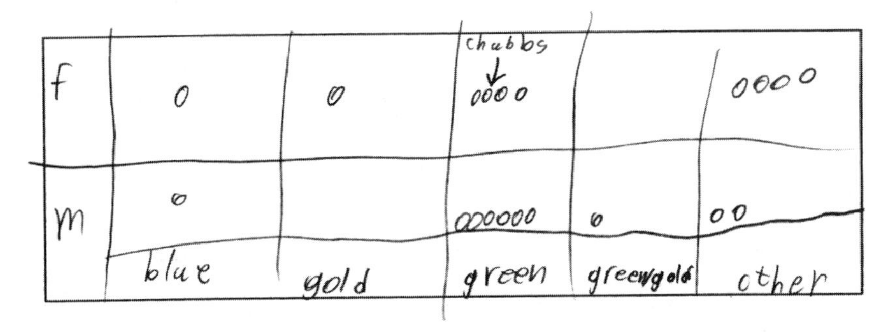

Comments: The student drew a graph of gender and eye color, which are not relevant to the question about Chubbs. Although the student correctly concludes that Chubbs is not chubby, he does not provide any reasons. It seems that the student is confused about how to use a graph and write reasons to support conclusions.

8. Here are all the nicknames and possible cats.

Nickname	Cat	Attributes	Your reasons
Baby Blue Eyes	Name: **Keikko** Case: **3**	***Age*** ***Eye_Color***	Justin: She's the youngest cat with blue eyes.
Stringbean	Name: **Beauportes** Case: **24** Name: **Semantha** Case: **18** Name: **Spud** Case: **38**	***Weight*** ***Body_Length***	Sample answer: **Beauportes is a good match to the nickname Stringbean because she is one of the longest cats (29 in.) and weighs only 9 lb.**
Mini Mouse	Name: **Fuzzy** Case: **5**	***Body_Length*** ***Weight***	Sample answer: **This cat is a good match to the nickname Mini Mouse because she is the smallest cat in terms of body length (13 in.) and weight (2 lb).**
Tall Tail	Name: **Spud** Case: **38**	***Tail_Length***	Eana: This cat is a good match for Tall Tail because she had the biggest tail length of them all.
Old Gold	Name: **Wiley** Case: **46**	***Age*** ***Eye_Color***	Sample answer: **This cat is a good match to the nickname Old Gold because she is the oldest cat with gold eyes.**

9. Make sure that students use evidence from the data to explain why their nickname is a good match to a particular cat. Sample student work:

   Thomas: I choose the nickname flashlight for the cat Sassy, case # 17. I think that this is a good nickname for this cat because her eyes are yellow and that is the color of a flash light.

   Comments: Student comes up with a creative nickname to match a cat's eye color (although Sassy's eyes are actually yellow/green).

   Justin: I choose the nickname Big Yellow for the cat Ravena, case # 43. I think that this is a good nickname for this cat because the cat is the heaviest and has yellow eyes.

   Comments: He comes up with a nickname to match two attributes: the cat's weight and eye color.

10. Mystery Attribute A: *Body_Length;* Mystery Attribute B: *Age;* Mystery Attribute C: *Weight*

11. Sample answer: I figured out that graph b was age because it went from 1 to 8 and those numbers are too small to be body lengths or weights.

Is the Cat Named Chubbs Chubby?

Name:

In this activity, you will compare an individual cat (Chubbs) to the other cats in his data set. Then you'll find nicknames for cats in a larger data set.

CHUBBS

1. Open the TinkerPlots file **One Year Old Cats.tp** to see data on 20 cats.

2. Find Chubbs using the information shown. Try to find Chubbs by making a plot, not by going through the data cards. How much does Chubbs weigh? Fill in the blank.

Attribute	Value	Unit
Name	Chubbs	
Gender	Male	
Weight		pounds
Body_Length	22	inches

Next you will analyze the data to find the answer to this question: Is Chubbs chubby compared to this group of 20 one-year-old cats?

3. Which attribute or attributes will you plot to answer the question?

   Name Gender Weight Body_Length Tail_Length Eye_Color

4. Make different plots using TinkerPlots to compare Chubbs to the group of cats.

5. Draw a copy of a plot that you found particularly helpful. Draw an arrow to show where Chubbs is on the plot.

6. Communicate your conclusions: Is Chubbs chubby compared to this group of cats? Write your answer on a separate sheet of paper.

NICKNAMES

7. Next you will be working with a larger data set of 50 female cats of different ages. Open the TinkerPlots file **Female Cats.tp.**

8. Pick two nicknames from the list. For each nickname, choose one cat that is a good match for this nickname. Tell which attributes you used and your reasons that this cat is a good match to the nickname.

   Baby Blue Eyes   Stringbean   Mini Mouse   Tall Tail   Old Gold

Nickname	Cat	Attributes	Your reasons
	Name ________ Case ________		
	Name ________ Case ________		

9. Come up with your own nickname for a cat. Write your answer on a separate sheet of paper.

   I choose the nickname ________ for the cat __________, case ________. I think that this is a good nickname for this cat because

MYSTERY ATTRIBUTES

10. Open the TinkerPlots file **Mystery Attributes 2.tp.** This file also contains the data on female cats. Which attribute does each line plot show? Figure out the mystery attribute, then make the graph with TinkerPlots.

    Mystery Attribute A ______________

    Mystery Attribute B ______________

    Mystery Attribute C ______________

11. Pick one graph. How did you figure out the mystery attribute?

OVERVIEW

This lesson is designed to build students' understanding of line plots and distributions. Line plots (also known as dot plots) are particularly useful representations for analyzing distributions. Line plots tend to be accessible for students to analyze because each case is visible on the graph as a separate dot. The lesson helps to prepare students for using line plots with TinkerPlots in Lessons 1.5, 1.6, 1.8, and 1.9. In addition, the work that students do in this section with line plots helps prepare them for working with other graphical representations, particularly histograms and box plots.

Objectives

- Interpret line plots
- Describe the shapes of distributions using informal terms
- Interpret what the shape of a distribution means about the data
- Compare the shapes of distributions
- Determine medians
- Analyze a larger data set of 50 cats

Offline

Class Time: One class period. If students are already familiar with line plots and medians, you can skip the Line Plots and Medians worksheets, or assign them as homework.

Materials

- Line Plots worksheet (one per student)
- Cats' Weights and Tail Lengths worksheet (one per student)
- Medians worksheet (one per student, on CD)
- Word Bank for Describing Data transparency (*optional,* on CD)
- Chart paper (*optional*)
- Sticky notes (*optional*)

LESSON PLAN

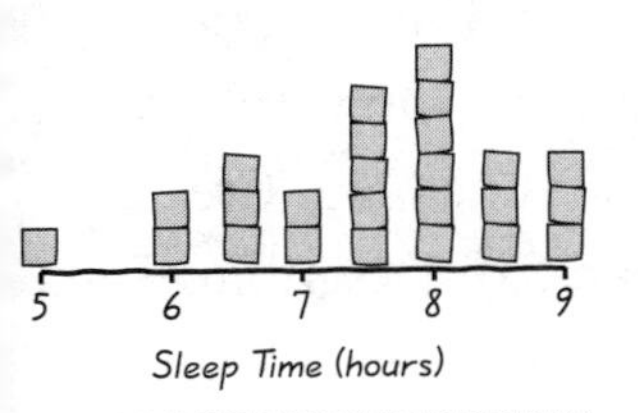

Save the class graph to use with the Medians activity later in the lesson.

Introduction

1. Make a class line plot to answer the question: "How many hours did you sleep last night?" Draw a number line for the horizontal axis on the board or on chart paper. Have students come up and mark an X (or place a sticky note) for their response. Discuss the class's results.
   - What's a typical amount of sleep for our class? How can you tell from the graph?
   - What do you notice about the shape of the data? What does this show about how much students sleep in our class?
   - What do you think the graph would look like for a class of first-graders? For twelfth-graders? Why? Ask students to make a sketch of what they think the shapes of those graphs would look like.
2. Use the class line plot to go over the characteristics of line plots:
   - There is one axis.
   - The axis is a number line. It need not start at 0.
   - The dots or X's are stacked above the values on the number line.
   - Each dot or X represents one case. In TinkerPlots, each dot represents one cat. For the class graph, each X or sticky note represents one student.
   - Line plots are also called *frequency graphs* and *dot plots.*

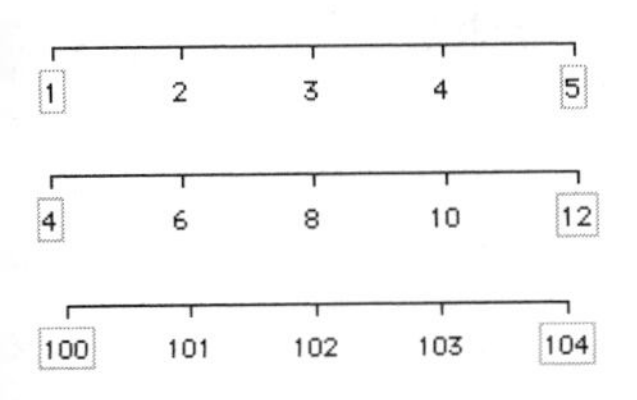

3. Have students practice reading information from a line plot using the Line Plots worksheet. Students use the line plot to figure out the number of cats with particular characteristics, such as a body length of 14 in. Students also need to figure out what percentage of cats has that characteristic.

   Note: Determining percentages was included in this lesson to prepare students for using percentages in Lesson 1.5, but it is not an essential part of this lesson. To meet your students' needs and your math goals, you might change the task to determining fractions, or cut this part of the activity.

Exploration

4. In the Cats' Weights and Tail Lengths activity, students compare two line plots. The focus is on comparing the shapes of the distributions. Introduce the term *distribution.* Have the class brainstorm words to describe the shapes of the distributions for the two line plots. Create a class word bank on chart paper, use the Word Bank for Describing Data transparency on the *Digging into Data* CD, or have students add their ideas to the word bank on the worksheet.

5. Go over how to fill in the table in question 2. This format will be used in many subsequent lessons. In the first column, students should write their observations about the shape of the distribution, such as: "The graph of tail lengths has a peak at the right end of the plot." In the second column, students should interpret and explain what this means about the data (cats), such as: "Many cats have similar tail lengths of 10–12 in."

   Have students work independently to look for similarities and differences in the shapes of the two distributions. They need to interpret what these differences mean about the cats.

Wrap-Up

6. Have a class discussion about the two line plots.
   - What similarities and differences do you see in the shapes of the two distributions?
   - Imagine that you are in a room with these 50 cats. What would you see? Would you see more variety in the weights of the cats or in their tail lengths? How can you tell from the line plots?

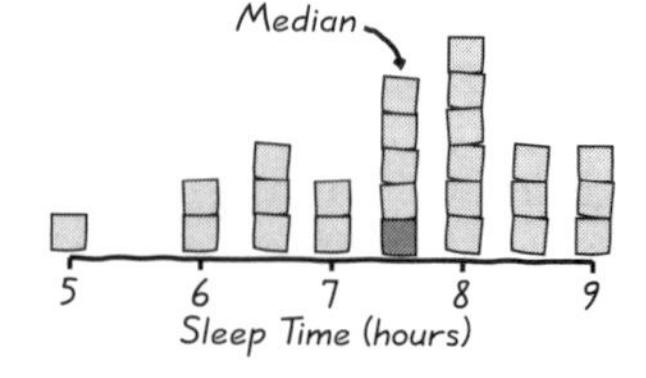

7. In the next lesson, students need to use medians. If students are unfamiliar with medians or need a review, you can use the class line plot of number of hours of sleep (from earlier in this lesson) to demonstrate the concept of median. If you created the line plot with sticky notes, then have two students come up and stand at each end of the graph. Have the students alternate removing one sticky note at a time from their side of the line plot—so one student starts by removing the sticky note with the largest value, and then the other student removes the sticky note with the smallest value. They continue until

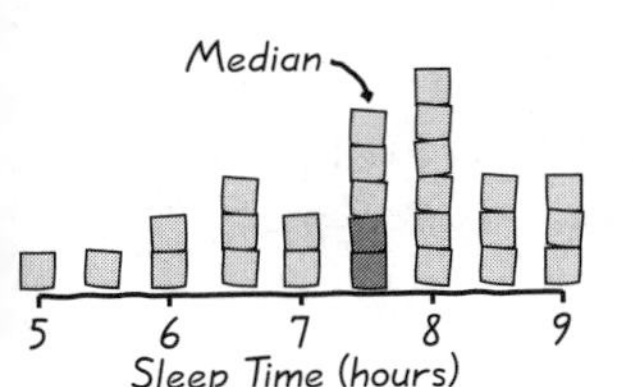

there is only one sticky note (value) left (if there was an odd number of values in the data set)—this is the median value. If there was an even number of values in the data set, then there will be two sticky notes left and the median is the average of those two values. If you created the line plot by drawing X's, the two students can alternate covering X's on each side with sticky notes until they are left with an X that is the median (or two X's that they need to average for the median).

8. If students need the review, have them complete the Medians worksheet. Students figure out the medians for two small data sets (questions 1 and 2). Then they create a data set that has values for 7 cats and a median of 11 lb (question 3).

9. Have a class discussion about finding medians.

   - How can you use a line plot to help you find the median?
   - How can you find medians when there are an odd number of values?
   - How can you find medians when there are an even number of values?
   - What strategies can you use to make up a data set with weights of 7 cats and a median of 11 lb?
   - A classmate tells you that the median weight for a data set of cats was 10 lb. Would it be correct to assume that there is at least one cat in that data set that weighs 10 lb? Why or why not?

ANSWERS

Line Plots

1. 5 cats; $\frac{5}{20} = 25\%$
2. 22 inches; 2 cats; $\frac{2}{20} = 10\%$
3. 13 inches; 1 cat; $\frac{1}{20} = 5\%$
4. 7 cats; $\frac{7}{20} = 35\%$
5. 4 cats; $\frac{4}{20} = 20\%$
6. Sample answer: Body length greater than 15 inches and less than 20 inches; 12 cats; $\frac{12}{20} = 60\%$

7. Sample answer: Tabby is one of the longest cats in this group of 20 cats. He is longer than 16 of the 20 cats, which is 80% of the group. Only 3 cats are longer than Tabby.

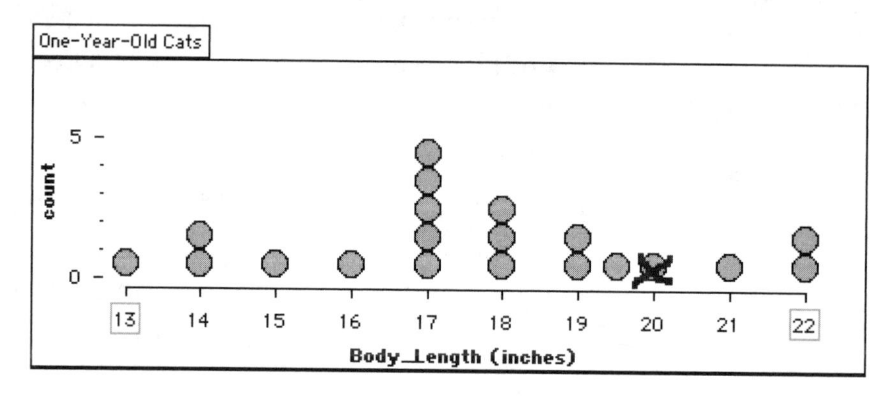

Cats' Weights and Tail Lengths

1. Sample answers:

   a. The shape looks like a hill with the peak in the middle of the graph.

   b. The data are bunched up on the right side of the graph. There is a large gap between 2 and 6.

2. Sample answers:

What similarities and differences do you see in the shapes of the two distributions?	What does this similarity or difference in shape tell you about the cats?
• **The tail lengths are bunched up on the right side of the plot, while the weights are more spread out.**	• **This means that there is less variety in the tail lengths and more variety in the weights.**
• **The peak of the weights is located in about the middle of the plot, whereas the peak for the tail lengths is located at almost the right end of the plot.**	• **The most common weights are in about the middle of the range, while the most common tail lengths are at the high end of the range.**

Sample student work (students were given question 2 without the table):

Yom: 1. The highest hill of weight is on center of graph but tail length is on left. 2. The dots of weight spread out but dots of tail length clustered on left of graph. 3. The dots of tail length have one big gab [gap] but the dots of weight don't have big gab [gap].

Comments: The student gives a thorough description of the shapes of the two distributions. This student considers the position of hills and whether the dots are spread out or clustered together, though the dots for tail length are actually clustered at the right, not the left, of the graph.

Jessica: The distribution for weight is fairly spread out, while the distribution for tail length was mostly on one side. The distribution for weight was fairly smooth, while the distribution for tail length was very "jagged" or had many ups and downs. The highest point for weight distribution was nearly in the middle, while it was almost on the end for tail length.

Comments: Her descriptions of the distributions focus on whether they are spread out and the position of the highest point. It's not clear what she means by the distribution for weight being "smooth" and tail length being "jagged."

Raina: 1. in the 1st one most of the cats are in the middle. 2. in the 2nd one most of the cats are towards the end. 3. 1 plot is weight 1 is tail length.

Comments: The first two statements focus on the location of most of the cats: middle versus end. The third statement, however, is not about shape.

3. Sample student work:

Yom: There is more variety in the cats' weight because the dots didn't cluster together but spread out fairly.

Comments: This student's answer is correct, though not very detailed.

Jessica: The distribution for the weights of the cats was spread out, meaning there were a variety of weights. The distribution in the tail length of the cats, however, was almost completely on one side, meaning the tail lengths for the cats were all very similar.

Comments: This student's strong response provides a detailed explanation of why the distribution for weight has more variety.

Raina: [More variety in the] weight [because] the range is bigger

Comments: This student's response shows a misconception. The range shows the distance between the lowest and highest values and is not a strong indicator of how spread out those values are within the range. In fact, the ranges for the two graphs are very similar: 12 and 11.

Medians

1. a. 7 lb
   b. Sample answer: Put the data in order and find the middle number.
2. a. $7\frac{1}{2}$ lb
   b. Sample answer: I averaged the two middle numbers.
3. a. Sample answer: 6, 6, 8, 11, 12, 12, 12
   b. One strategy is to draw 7 blank lines for the weights of the 7 cats. Then, because there are an odd number of values, put the median value of 11 in the middle blank. Choose 3 numbers less than or equal to 11 to fill in the first 3 blanks, and 3 numbers greater than or equal to 11 for the last 3 blanks.
   c. One strategy is to add 1 cat on either side of the median. Another is to add both cats at the median.
   d. One strategy is to add 1 cat on either side of the median and 1 cat at the median.

Line Plots

Name:

You will be using line plots to analyze data about cats. Here are some characteristics of line plots.

- There is one axis.
- The axis is a number line. It does not need to start at 0.
- The dots are stacked above the values on the number line.
- Each dot represents one case. Here, each dot represents one cat.

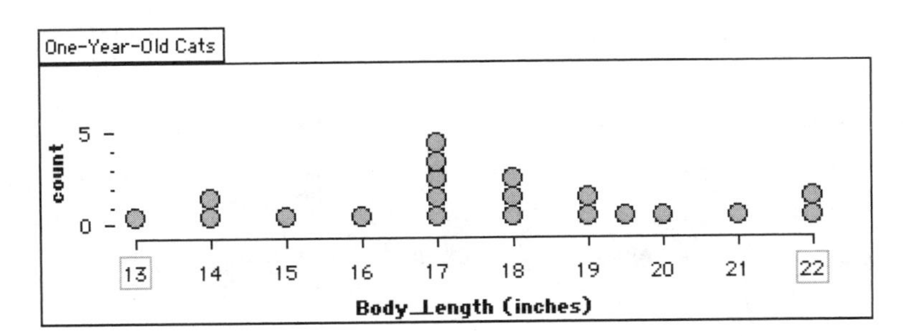

Read the line plot to find cats with each characteristic, then tell how many cats have that characteristic and what percentage they are of the total number of cats. For question 6, add your own characteristic.

Characteristic	Number of cats	Percentage of cats
Body length of 14 inches	*2*	$\frac{2}{20} = \frac{10}{100} = 10\%$
1. Body length of 17 inches		
2. Longest body length: ________ inches		
3. Shortest body length: ________ inches		
4. Body length greater than 18 inches		
5. Body length less than 16 inches		
6.		

7. Put an X on the dot that represents Tabby Burton, who is 20 inches long. How does Tabby Burton's body length compare to this group of one-year-old cats? Write your answer on a separate sheet of paper.

Cats' Weights and Tail Lengths

Name:

In this activity, you will compare the distribution of 50 female cats' weights with the distribution of their tail lengths.

1. What words would you use to describe the shape of each distribution? Choose words from the Word Bank below and write them on each line plot.

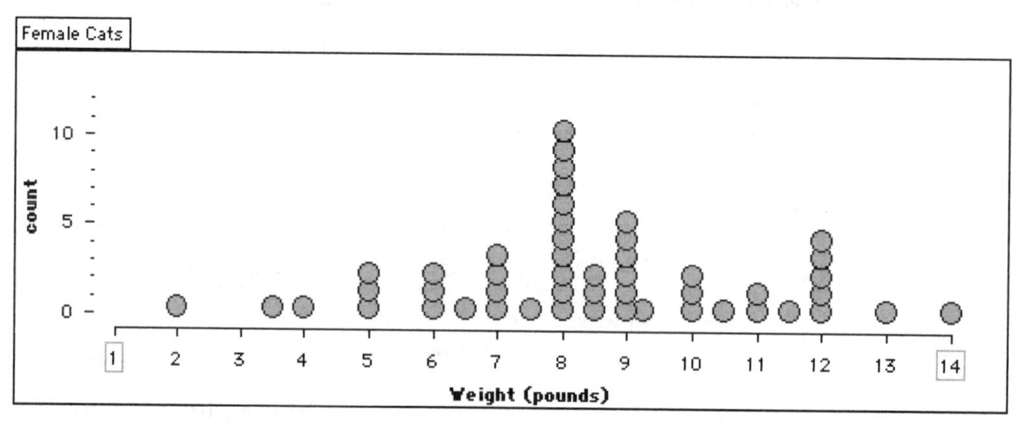

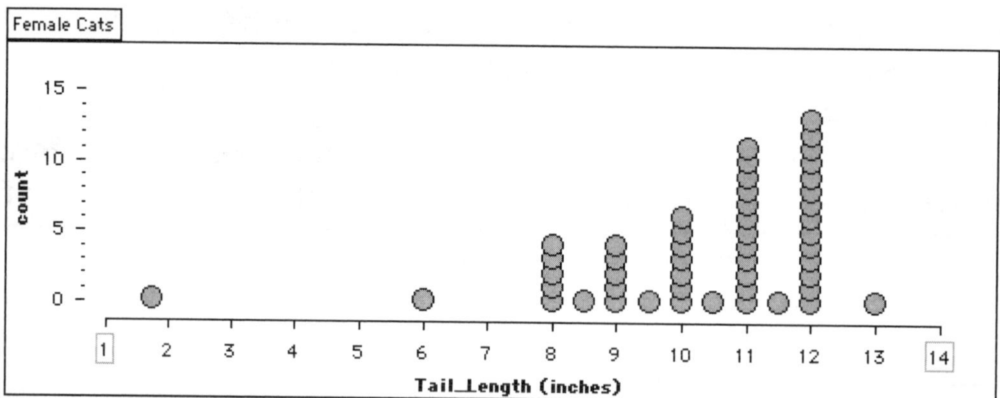

WORD BANK

Here are some words that you can use to describe data and how they appear on graphs. Add your own ideas to the list.

Spread out Bunched together Skewed Bunched up on one side Symmetrical Mirror-image Center Clump Cluster Flat Stairs Bell-shaped Ski jump Peak Hill Valley Gap Hole

2. How do the shapes of the two distributions compare?

What similarities and differences do you see in the shapes of the two distributions?	What does each similarity or difference in shape tell you about the cats?

3. Imagine that you are in a room with this group of cats. Would you expect to see more variety in how much the cats weigh or in how long their tails are? Why?

Using Center Clumps

OVERVIEW

This lesson is designed to help students begin making the transition from focusing on individual cases (cats) to thinking about the group. In previous lessons, students focused on maximum and minimum values, which are the ends of the distribution. In this lesson, students need to shift their focus to the center of the distribution to figure out what's typical for the group. The dividers feature in TinkerPlots provides a visual way for students to find a "center clump." By using the dividers and percentages, students can experiment with changing the positions of the dividers to find the middle 50% of the data. This thinking also lays groundwork for the use of box plots. It can be challenging for students to view the data as a group; for that reason, they will have the opportunity to continue working on this concept in subsequent lessons in this section and in Sections 2 and 3.

Objectives

- Begin making the transition from looking at individual cases to thinking about the group
- Experiment with different ways to determine what is typical for the group of cats
- Deepen understanding of the meaning of the term *typical*
- Start using center clumps to identify what's typical for a group
- Use percentages to find the middle 50% of a group

TinkerPlots

Class Time: One class period

Materials

- What Is Typical for This Group of Cats? worksheet (one per student)

Data Sets: Male and Female Cats.tp (100 male and female cats), **One Year Old Cats.tp** (20 one-year-old cats) for demonstration

TinkerPlots Prerequisites: Students should be familiar with basic graphing and showing medians.

TinkerPlots Skills: Using dividers and showing percentages are explained in this lesson.

LESSON PLAN

Introduction

1. Explain that in this activity, students need to shift from looking for individual cats that "stand out" to describing the group of cats. This involves making a transition from finding individual cases to answer questions such as: Who is the longest male cat?, to looking at the whole group of cats as an aggregate to answer questions such as: How would you describe the body lengths for the group of 100 cats?

Exploration

2. Students use TinkerPlots to complete question 2. They can apply what they've learned about distributions to describe the shape of the data. It might be helpful to post a class word bank of terms, or for students to use their word bank from Lesson 1.4.

3. The focus of the rest of the lesson is on determining what are typical body lengths for the group of cats. It's important to stress that what's "typical" or "average" is a characteristic of a group and not of an individual. Encourage students to look for a range of typical body lengths, or a numeric measure, instead of picking one cat and saying that cat has a typical body length.

4. Show students a line plot of *Body_Length* and ask them what are typical body lengths for the group of cats. Some students may think of the mode as representative of what's typical, and say that the typical body length is 17 in. Other students may select a range of body lengths, such as 17 to 19 in. Do not try to have the class reach a consensus. Students will refine their ideas of what's typical for the group as they go through the subsequent activities.

5. Introduce how to use dividers in TinkerPlots. (The directions are on the worksheet.) You can demonstrate question 3 using **One Year Old Cats.tp.** Discuss this question, and point out that students will be working with a different data set. Encourage students to move the dividers around, and to use the gray area to show what they think are typical body lengths for the group of 100 cats. Encourage them to try to find a *center clump* of the data.

   • Where did you place the dividers to show the typical body lengths for the group of cats? Why?

6. Explain how to use percentages in TinkerPlots. (Select the plot and click the **%** button.) When students move the dividers, the percentages change, and this can help them decide where to position the dividers to show what's typical. Then have students look at question 5, which shows two examples of different positions for the dividers. Students need to decide which example uses the gray area to best show the typical body lengths for the group of one-year-old cats. Discuss the question, and the meaning of the percentages. Next, students need to use percentages to position the dividers themselves to show typical body lengths for the group of 100 cats. If students are having trouble reading precise values, encourage them to use reference lines. (Select the plot and click the **Ref.** button.)

Wrap-Up

7. Discuss students' answers to questions 6 and 7.

   - What percentage of values did you put in the gray area? Why?
   - Where did you position the gray area? Why?
   - What are typical body lengths for this group of 100 cats?

   The discussion should lead to some agreement about using the middle 50% of the data to represent what's typical for the group. Depending on the distribution of the data, it's not always possible to get exactly 50% in the gray area and 25% on either side. Students should try to use the dividers to enclose approximately 50% of the data in the middle, with approximately equal amounts of values on either side.

8. Introduce the use of the word *tend* to describe what's typical for a group. For example, basketball players *tend* to be between 6 and 7 ft tall. The word *tend* means that you are describing what's typical, and that there can be exceptions—such as basketball players like Kareem Abdul-Jabbar, Shaquille O'Neal, and Yao Ming, who are over 7 ft tall. In this lesson, we can say that these 100 cats *tend* to have body lengths from 17 to 21 in.

ANSWERS

What Is Typical for This Group of Cats?

2. Sample answer: The distribution is shaped like a broad hill toward the left of the graph. This means that cats can have any of these body lengths, but are more likely to have the shorter lengths.

3. Trevor's gray area has more values in it while being the same width as Gabriella's, so it better shows what's typical for the group of cats.

4. Here are some examples of where students may want to place the dividers.

   This student has enclosed a large area of values to show that he thinks typical body lengths are from 15 to 23 in. He has chosen the body lengths that have five or more values.

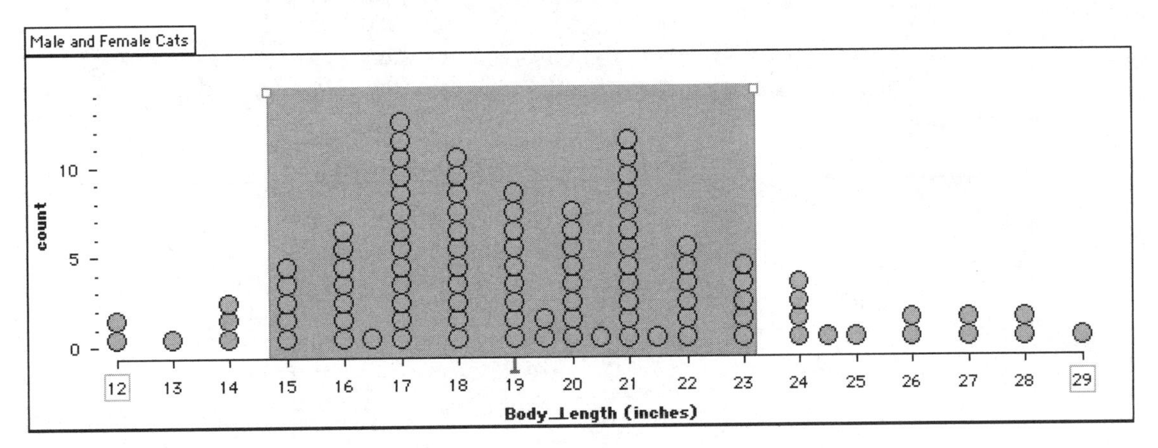

Here the median is in the middle of the gray area and shows that typical body lengths are from 17 to 21 in.

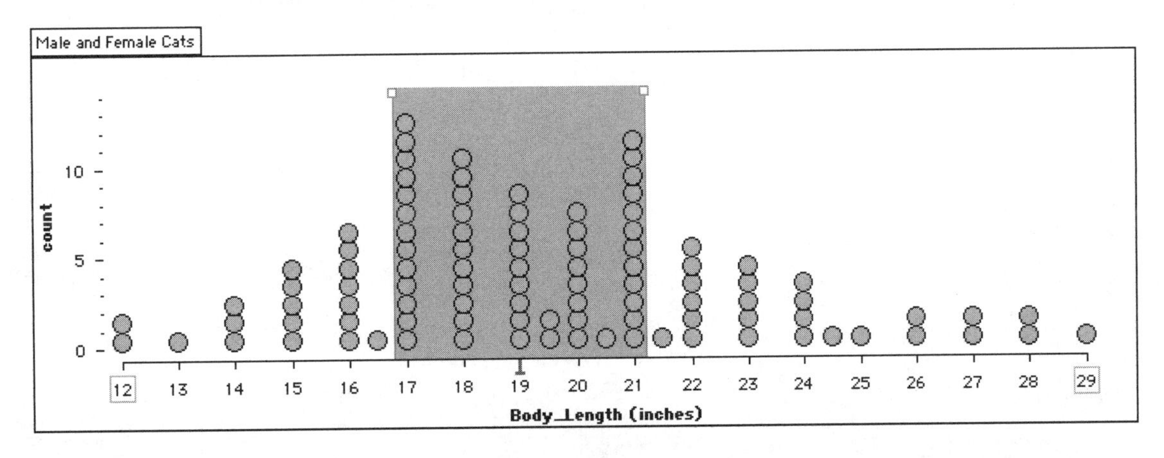

5. Bonita has used the dividers to show the middle 50% of the data. It's in the middle because 25% of the values are to the left and 25% are to the right.

6., 7. This graph shows the gray area with about 50% of the values positioned in approximately the middle of the data. Due to the distribution of the data set, it is not possible to position the dividers to have three sections with exactly 25%, 50%, and 25%. Student writing will vary, but should include a range of approximately 50% of the values.

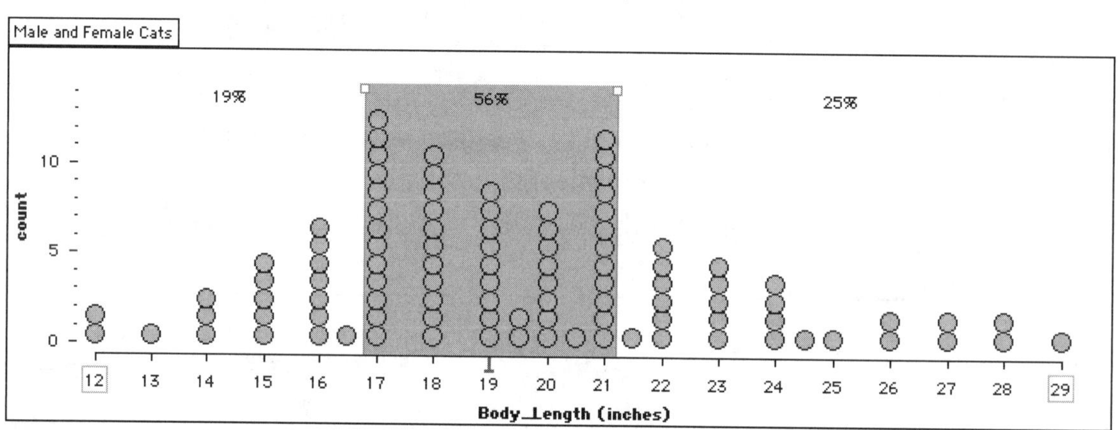

What Is Typical for This Group of Cats?

Name:

You will use *center clumps* to determine what is typical for a group of 100 cats.

1. Open the TinkerPlots file **Male and Female Cats.tp** to see data on 100 cats.

2. Make a line plot to see the distribution of cats' body lengths. What does the shape of the distribution tell you about the group of cats?

You need to find *typical* body lengths for these cats. You can use *dividers* in TinkerPlots to show the typical body lengths of cats on your line plot.

3. Which student used the gray area to best show the typical body lengths for this group of one-year-old cats? Why?

Gabriella

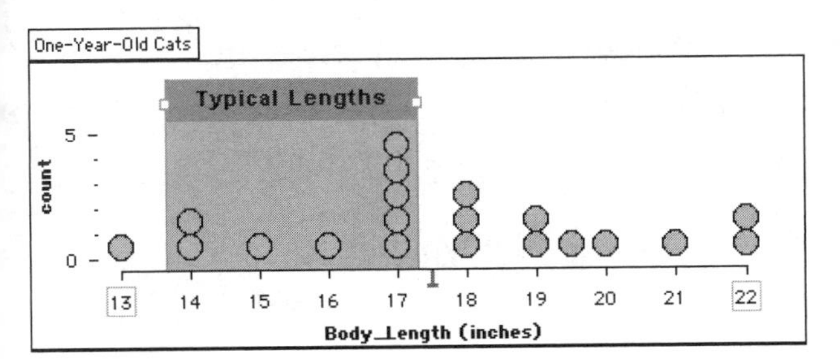

Trevor

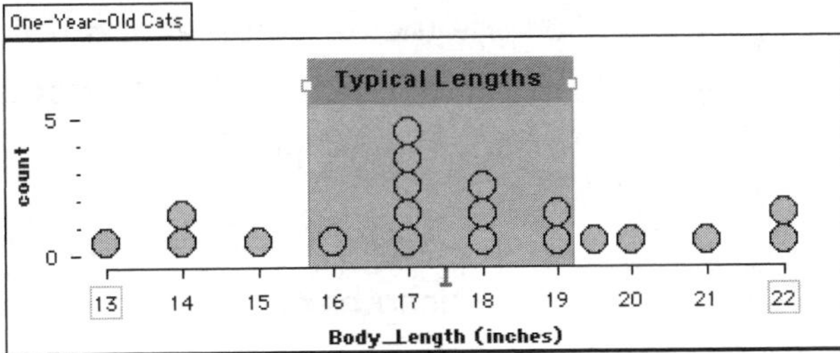

4. Use the dividers to show typical body lengths for the group of 100 cats. To add dividers, select the plot and click the **Div.** button. Experiment with moving the dividers around the plot. (Drag the white box.)

   Where did you position the dividers to show the typical body lengths of cats in this group?

   ________ in. – ________ in.

You can use the dividers, the median, and percentages to find a *center clump* of values that represents what is typical.

5. Which student used the gray area to best show the typical body lengths for this group of one-year-old cats? Why?

Andrew

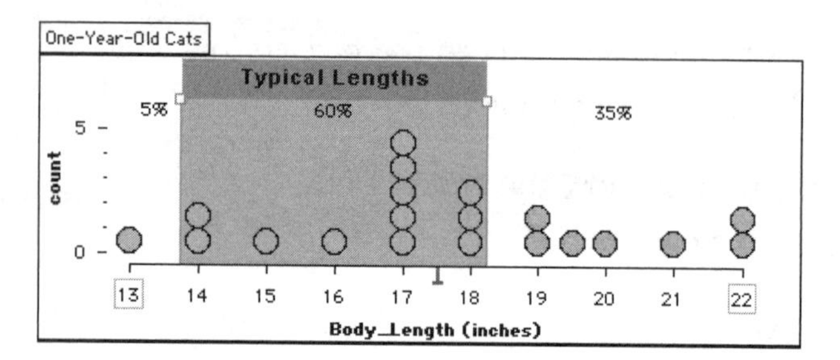

Bonita

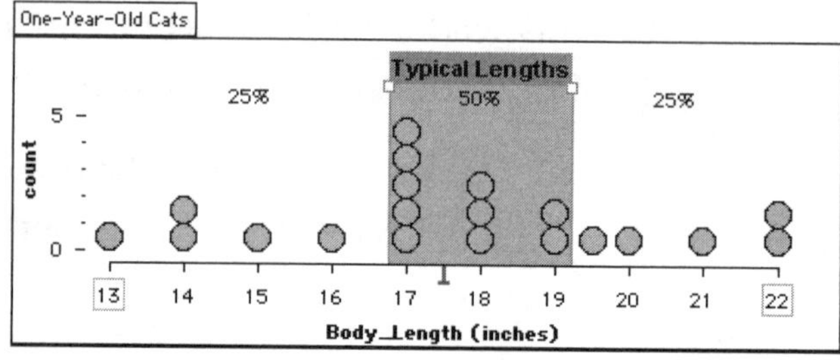

6. Use the percentages and the median to show a clump that represents the typical body lengths for this group of 100 cats. Use a reference line to read precise values. (Select the plot and click the **Ref.** button.)

   a. Where did you position the dividers to show the typical body lengths of cats in this group?

      ________ in. – ________ in.

   b. What percentage of cats is in the gray area? ________

   c. What percentage of cats is in each of the white areas? ________

7. Write at least two sentences on your own paper about your findings for the question: What are typical body lengths for this group of 100 cats? Make sure to use evidence from the data to support your answer.

OVERVIEW

Students use TinkerPlots to compare a group of male cats to a group of female cats, to determine which group tends to weigh more. This lesson has a lot of structure to help students build their repertoire of ways to make comparisons and to address the tendency of some students to compare data in just one way. The lesson guides students to try a variety of methods for making comparisons and to recognize that this will give them a fuller picture of the data. In later lessons, students will have the opportunity to apply these methods in more open-ended investigations.

If you would prefer to use a less structured approach, have students use the Investigating Data worksheet from the *Digging into Data* CD. If students do not come up with the methods in the lesson, then you can introduce them later.

Objectives

- Use a variety of ways to compare two groups, including comparing shapes, measures of center (means and medians), and measures of spread (ranges) (This lesson assumes that students have had prior experience with these measures.)
- Recognize the importance of analyzing data in different ways to make more comprehensive comparisons

TinkerPlots

Class Time: One class period. Students will write conclusions in Lesson 1.7.

Materials

- Weights of Male and Female Cats worksheet (one per student) or Investigating Data worksheet (one per student, on CD in the file **Templates.pdf**)

Data Set: Male and Female Cats.tp (100 male and female cats)

TinkerPlots Prerequisites: Students should be familiar with basic graphing.

TinkerPlots Skills: Using the **Drawing** tool and dividers, and showing counts and percentages are explained in this lesson.

LESSON PLAN

Introduction

1. Introduce the question students will be investigating: How do the weights of male and female cats compare? Ask students to make hypotheses individually, and then discuss with the class. Students come up with a variety of reasons for which gender they think will weigh more (question 1).

2. Before students begin analyzing the data, demonstrate how to use the **Drawing** tool and how to clear drawings. This tool will help students compare the shapes of the distributions. Click the **Drawing** tool, then click and drag in the plot to draw.

   You need to click the button for each new drawing. To clear all drawings, choose **Clear Drawing** from the **Icon Size** menu (to the right of the **Drawing** tool). Encourage students to use the **Drawing** tool to sketch the overall shape of the data, rather than going up and down for every dot.

Exploration

3. Students use the **Drawing** tool to compare the shapes of the distributions (questions 4 and 5).

4. Next, students need to compare typical weights for the group of male cats with typical weights for the group of female cats. By applying what they learned in Lesson 1.5, they can use dividers, percentages, and medians to find a center clump for each group (questions 7–9).

   The graph on the next page shows one way to position the dividers. However, students may want to position the dividers in different places. Because of the way the data is distributed, it's not possible to divide it into three parts that are exactly 25%, 50%, and 25%. Encourage students to place the dividers so that the gray area shows approximately the middle 50% of the values.

   Students need to compare the positions of the center clumps. The center clump for the males is to the right of the clump for the females, which shows that the typical weights for males are higher than the typical weights for females.

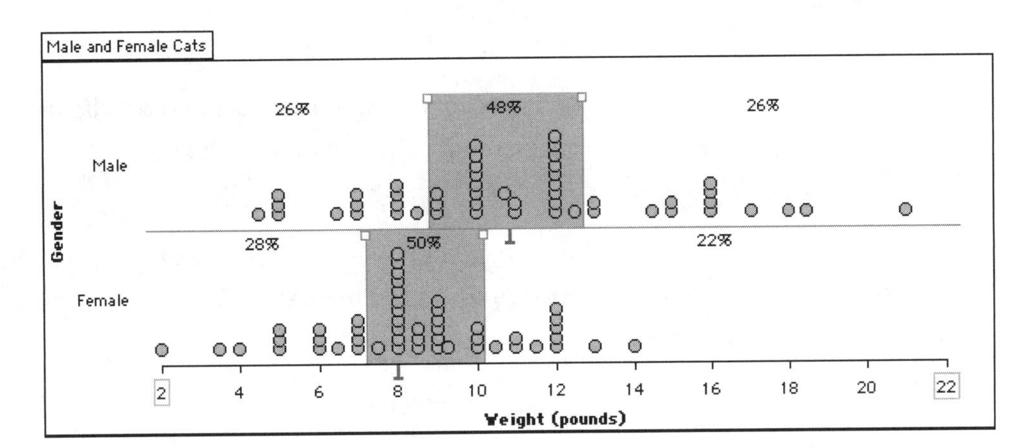

5. In the last part of the comparison, students use the median, mean, and range to compare the groups of male and female cats.

Wrap-Up

6. Discuss students' findings about the two groups of cats.
    - How do the shapes of the distributions compare for the weights of male cats and female cats?
    - How do the typical weights for the group of male cats compare with the typical weights for the group of female cats?
    - How do the measures of center and the range compare?
    - Why is it important to look at both center and spread?
    - What other questions would you like to investigate about cats? How would you collect the data?

ANSWERS

Weights of Male and Female Cats

1. Sample student hypotheses:

   Females tend to be heavier.

   Huan-Jie: I think that female cats are going to weigh more because other animals females are usually heavier than males.

   Justin: I think that female cats tend to weigh more because they are the gender that have babies.

   Katy: I think that female cats tend to weigh more because they are less energetic.

Males tend to be heavier.

Adalia: I think that the males are going to be heavier because they tend to be more lazy. Also I have a brother and sister and the male is 8 pound heavier.

Yom: I think that male cats tend to weigh more than female cats because usually males eat more than females and are bigger than females.

Thomas: I think that male cats will weigh more because males are usually bigger (any species)

5. Sample answer:

What similarities or differences do you see in the shapes?	What does this similarity or difference tell you about how the weights of male cats compare to the weights of female cats?
• **Both distributions are hill-shaped and fairly symmetrical.**	• **It is more common for both male and female cats to have weights in the middle values—neither the low values nor the high values are very common.**
• **The distribution for males is much broader and shorter, while the females are more clustered together.**	• **The weights of the male cats vary more than the weights of the female cats.**

8. Sample answer (placing the dividers as in part 4 of the lesson plan): Females: 7–10 lb; males: 8–13 lb

9. Sample answer: Male weights are typically greater than female weights. The center clump for males is to the right of the center clump for females.

10. Median: 10.9; 8.0; Males' median weight is higher by 2.9 lb.

    Mean: 11.1; 8.4; Males' mean weight is higher by 2.7 lb.

    Range: 16.5; 12; There is a larger difference between the highest and lowest weights for male cats than for female cats.

11. Answers will vary.

Weights of Male and Female Cats

Name:

In this activity, you will compare two groups by making visual comparisons of the shapes of the distributions and by making numeric comparisons.

ASK A QUESTION AND MAKE A HYPOTHESIS

You will investigate the question: How do the weights of male and female cats compare?

1. What do you think the data will show? What are your reasons for this hypothesis?

ANALYZE DATA

You will graph the data in different ways to compare the weights of male and female cats. You will write your conclusion in the next lesson.

2. Open the TinkerPlots file **Male and Female Cats.tp** to see data on 100 cats (50 male and 50 female) of different ages.

Compare the Shapes of the Distributions

3. Make a line plot of the cats' weights and then separate by *Gender* to show the males and females.

4. Use the **Drawing** tool to sketch the overall shape of the data for males' weights and females' weights. This graph shows the shape of the data for the female cats.

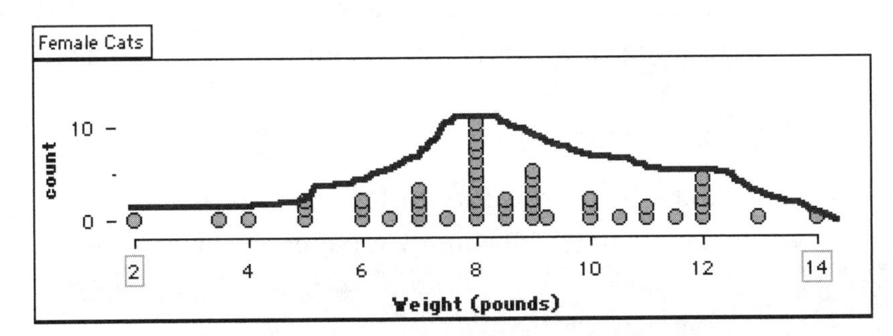

5. Compare the shapes of the distributions of the weights of the male and female cats. Use at least two words from the word bank.

What similarities or differences do you see in the shapes?	What does this similarity or difference tell you about how the weights of male cats compare to the weights of female cats?

Word Bank

Spread out **Skewed** **Symmetrical** **Center** **Clump** **Peak** **Gap** **Flat** **Bunched together** **Bunched up on one side** **Mirror image** **Cluster** **Valley** **Hill** **Hole** **Stairs**

Compare Typical Weights

6. Clear your drawings from the line plots by choosing **Clear Drawing** from the **Icon Size** menu (to the right of the **Drawing** tool).

7. Use the median, percentages, and dividers to figure out the typical weights for each group of cats. (Use reference lines to read precise values.)

8. Experiment with different ways to place the dividers. Then, decide what you think are typical weights for males and females.

   Females _________ lb – _________ lb Males _________ lb – _________ lb

9. How do the typical weights for females compare with the typical weights for males? Use your own paper for your answer.

Make Numeric Comparisons

10. How do these measures of center and spread compare for male and female cats' weights? Use TinkerPlots to fill in the table.

Measure	Male cats (lb)	Female cats (lb)	What does this tell you about how the weights of male and female cats compare?
Median			
Mean			
Range			

11. Experiment with your own ways to compare the weights of male and female cats. Keep notes on your findings.

Writing High-Quality Conclusions

OVERVIEW

Some students struggle with writing conclusions about data because they are unsure about what is expected. This lesson provides a rubric with clear expectations that can help guide students in writing their own conclusions. By first using the rubric to rate student work samples, students build an understanding of the different criteria in the rubric. The work samples are taken from students who are not in the class so that students will feel more comfortable rating them. There are many weaknesses in the samples, which gives students the opportunity to come up with many suggestions for improvement. The process of coming up with these suggestions will help students when they write their own conclusions.

Objectives

- Become familiar with a rubric for writing high-quality conclusions
- Build understanding of the characteristics of high-quality conclusions by rating work samples on a rubric
- Write high-quality conclusions that use the data to compare two groups

Offline

Class Time: One class period. Students can continue writing or revising their conclusions for homework.

Materials

- Characteristics of High-Quality Conclusions worksheet (one per student)
- Rubric for Weights of Male and Female Cats (two per student or pair)
- Rate the Conclusions worksheet (one per student or pair)
- Talking Math worksheet (one per student, *optional,* on CD)

LESSON PLAN

Introduction

1. Introduce the goal of the lesson: to prepare students for writing high-quality conclusions. They will be writing their conclusions about how male and female cats' weights compare using their analysis of the data from Lesson 1.6. Go over the rubric by giving some examples of the different criteria. If your students are unfamiliar with rubrics, you might want to reduce the number of criteria and focus on just two or three as a starting point.

Exploration

2. Hand out the student work samples and copies of the rubric. Have students work in pairs to rate each sample using the rubric. Point out that these work samples are from students who are having difficulty writing conclusions. The samples do not include a high-quality response because students might be tempted to copy it when they write their own conclusions. (Examples of strong responses are given in the answers.)

3. Discuss the ratings for each conclusion for each characteristic of the rubric.

   - How did you rate each conclusion? Why? Give specific examples from the work to support your rating.
   - What suggestions would you give to each student to improve his or her conclusion?
   - What criteria would you add to the rubric?

   The goal of discussing the ratings is to build students' understanding of what each rubric characteristic means. There is bound to be some disagreement about how to rate each piece of student work. It is not necessary for the class to come to a consensus.

4. Students write their own conclusion for the question: How do the weights of male and female cats compare? To help students get started, suggest that they read through their findings from Lesson 1.6 and star (*) the information that will be helpful for writing strong conclusions.

Wrap-Up

5. After students have written drafts of their conclusions, have them exchange their drafts with their partners and give each other feedback (using the rubric) to improve the conclusions. Alternatively, students can rate their own work on the rubric. Then students need to revise their own conclusions to better meet the criteria for high-quality conclusions. If students need additional practice with using mathematics terms, have them complete the Talking Math worksheet from the *Digging into Data* CD.

ANSWERS

Characteristics of High-Quality Conclusions

1., 2. Ratings will vary, but should be roughly similar to these.

Comments for Student X's conclusion:

Strong and specific evidence: 2. The first statement is not based on the data set. The evidence in the second statement is weak because it focuses on two individual cats and not on the groups of cats. The third statement needs to provide more specific evidence about the type of average and the values for the averages.

More than one kind of evidence: 2. The student compares the averages and the percentages of male and female cats that weigh less than 14 lb and more than 14 lb.

Clear and easy-to-follow conclusions: 2. The student does not say what kind of "average" he/she is comparing. The student does not compare the shapes of the distributions.

Compares groups: 2. The last two statements compare the groups of cats, but the second statement focuses on two individual cats—the biggest and the smallest.

Explains comparisons in context: 4. The student gives evidence in the context of cats' weights by using the units "lbs" and the words "cats," "weigh more," "biggest," and "smallest."

Total score: 12 out of 20

Suggestions for improvement:

Eana: *Need to add more comparison and more detail on why males weigh more than females.*

Ian: *Some info was not from the graph and you should take [talk] about the shape of the graph and the mode and median.*

Comments for Student Y's conclusion:

Strong and specific evidence: 2. The student compares the shapes of the distributions, but doesn't explain why being "spread out" instead of "clumped up" means that the males weigh more. It's not necessarily true that being "clumped up" means that females weigh less—they could be clumped up in the high weights. The student compares the modes, medians, and means but does not give specific values for these measures or say how much difference there is for the two groups.

More than one kind of evidence: 2. The student compares the shapes of the data and measures of center.

Clear and easy-to-follow conclusions: 2. The student does not explain what the differences in the shapes of the distributions show about the two groups. The statements need more explanation and specific evidence.

Compares groups: 3. The student focuses on comparing groups by using the shapes and the measures of center. The student does not compare any individual cats.

Explains comparisons in context: 2. The student gives evidence in the context of cats' weights by using the words "cats," "weigh more," and "weight." However, the student doesn't use the unit (pounds).

Total score: 11 out of 20

Suggestions for improvement:

Give more specific evidence by giving values for the means, medians, and modes and for how much higher the males' values were than the females'. Explain how the differences in the shapes give evidence that males weigh more.

3. Sample student work:

Kiara: My conclusion is that male cats weigh more than female cats. Firstly, they (males) have a higher median. Theirs is 10.875 while females' median is 8. Also, the male cats have a higher mean. Males is 11.135 and females have 8.445. Comparing the shape of the data, the males' peak was around the center, but the females peak was skewed towards the left. The males are distributed more widely and have a range of 16.5 whereas the range of females is only 12. Also, there were more male cats on the heavier side. Twenty-three males were over 12 lbs but only 7 females were over 12 lbs. All of these details conclude that males weigh more than females.

Comments:

Strong and specific evidence: 4. The student provides detailed support for the conclusions by giving specific measures and descriptions of differences in the shapes of distributions.

More than one kind of evidence: 4. The student compares medians, means, shapes of distributions, and numbers of cats over 12 lb.

Clear and easy-to-follow conclusions: 3. The sequence of statements is well organized. The student could make the conclusions clearer by giving units (pounds) for the means and medians, and by explaining what the differences in the shapes of the distributions mean.

Compares groups: 4. The comparisons focus on the two groups and not on individual cats.

Explains comparisons in context: 3. This could be strengthened by giving units (pounds) for the measures and explaining that if the median weight for males is about 3 lb more than females, this is a large difference in weights given the weights of cats.

Total score: 18 out of 20

Eana: My conclusion is that males tend to weigh more than females. In the chart the heaviest male weighed 21 lbs and for females the heaviest weighed 14 lbs. The males were more spread out and the females were sort of bunched together on a side. The mean, median, mode, and range of the male is greater than the female. The mean of the male is 11.135 and the female is 8.445. The median of the male is 10.875 and for females it's 8. For mode it is 12 or [for] males and 8 for females. There are 25 female cats that weigh between 7 and 9 pounds.

Comments:

Strong and specific evidence: 3. The student gives specific values for measures.

More than one kind of evidence: 3. The student compares medians, means, modes, and shapes of distributions. It seems that she listed all the kinds of evidence she could think of. She could improve her conclusion by selecting the stronger pieces of evidence and explaining what they mean.

Clear and easy-to-follow conclusions: 3. The student could make the conclusions clearer by explaining what the differences in the shapes of the distributions mean and by giving units (pounds) for the measures of center.

Compares groups: 3. She compares the groups, but also compares individual cases: the heaviest cats.

Explains comparisons in context: 3. This could be strengthened by giving units (pounds) for the measures, and explaining what the differences in measures mean in the context of the weights of cats.

Total score: 15 out of 20

Thomas: My conclusion is that male cats tend to weigh more than female cats. By looking at my data I have noticed many examples of how male cats weigh more than female cats. One example I have is the avreges. Male cats have an avrage weight of 11 pounds and female have an average of 7 pounds. Another example I have is 21 pounds is the heaviest male cat and 14 is the heviest female cat. The last example I have is that the lightest male cat is 5 pounds and the lightest female cat is 2 pounds. This is how I have come to the conclusion that male cats tend to weigh more than female cats.

Comments:

Strong and specific evidence: 2. The student uses averages to support conclusions, but more evidence is needed.

More than one kind of evidence: 2. The student compares the averages. Comparing the heaviest cats and lightest cats is the same kind of comparison. The student does not compare the shapes of the distributions.

Clear and easy-to-follow conclusions: 3. The ideas are organized and easy to follow. The student doesn't say what kind of "average" he used.

Compares groups: 2. Two pieces of evidence are based on comparing individual cats.

Explains comparisons in context: 4. The student uses units (pounds) for all the values and uses "heaviest" and "lightest" to connect to the context of comparing weights.

Total score: 13 out of 20

Characteristics of High-Quality Conclusions

Name:

In this lesson, you will use a rubric to rate example conclusions and write suggestions for improvement. This will help you write high-quality conclusions.

1. Read the Rubric for Weights of Male and Female Cats. At the end of this lesson, you will use the rubric to write your own conclusions for the data about cats' weights in Lesson 1.6.

2. Use the rubric to rate Student X and Student Y's conclusions. These conclusions were written by two students from another class. Both students need help to improve their work.

   a. Read each conclusion.

   b. Use the rubric to rate the conclusion.

   c. Write suggestions on the rubric to help each student improve his or her conclusion.

3. Use what you have learned from rating the examples to write your conclusion for the question: How do the weights of male and female cats compare?

   a. Before writing, read the rubric for making comparisons, and read your list of findings from analyzing the data on cats' weights in Lesson 1.6. Star (*) the comparisons that you think provide the strongest support for your conclusion.

   b. Write your conclusion on a separate sheet of paper.

Rubric for Weights of Male and Female Cats

Name:

High quality					Needs improvement
Gives strong and specific evidence from the data to support conclusions.	4	3	2	1	Does not give evidence to support conclusions.
Gives more than one kind of evidence, such as comparisons of shapes, centers, and numeric measures. Selects evidence that provides strong support for conclusions.	4	3	2	1	Gives only one kind of evidence; for example, makes only numeric comparisons.
Conclusions are clear and easy to follow. Uses appropriate terms and units.	4	3	2	1	Conclusions are confusing and hard to understand. Does not use appropriate terms and units.
Compares the group of males to the group of females.	4	3	2	1	Does not compare the group of males to the group of females. Compares only individual cats; for example, "The heaviest female cat is 14 lb and the heaviest male is 22 lb."
Explains what the comparisons mean in the context of cats.	4	3	2	1	Makes comparisons, but does not explain what they mean in the context of cats.

Total rubric score: __________ points out of 20

Suggestions for Improvement

Rate the Conclusions

Name:

Two students wrote conclusions for the question: How do the weights of male and female cats compare? Use the rubric to rate their work and then write suggestions to help them improve.

STUDENT X'S CONCLUSION

My conclusion is that males weigh more then females.

-in most every other species the average male weighs more then the average females

-That the biggest cat on the chart is a male and the smallest cat is a female.

-Males average was higher then the females average.

-100% of female cats weigh 14 or less lbs and only 78% males weigh 14 and down [or less] the other 22% weigh more.

STUDENT Y'S CONCLUSION

My conclusion is That the males weigh more than the female cats. How come I say this is because when I looked at the data I saw that the weight of the females were more clumped up than the males. The males were spread out more. The mode, median and mean weight were all in favor of the male weight.

Working with Ratios

OVERVIEW

Students look at the ratio of a cat's tail length to its body length. They first find the ratios for individual cats, then use a formula to find the ratio for all the cats, and then use this ratio to come up with recommendations for drawing cats with realistic proportions. The concept of ratio is central to the middle-school mathematics curriculum. This lesson provides an opportunity for students to apply their knowledge of ratio to a data analysis context.

Objectives

- Use ratios to make comparisons
- Interpret and visualize the meaning of ratios; for example, what does a cat look like with a specific *Tail_Length* to *Body_Length* ratio? What does a low ratio mean in this context?
- Compare ratios
- Use a formula to create a new attribute that will be helpful for making comparisons

TinkerPlots

Class Time: One class period

Materials

- Tail Length and Body Length worksheet (one per student)

Data Set: Male and Female Cats.tp (data for 100 cats)

TinkerPlots Prerequisites: Students should be familiar with intermediate graphing.

TinkerPlots Skills: Creating a new attribute using a formula is explained in this lesson.

LESSON PLAN

Introduction

1. Introduce the context for the investigation: Artists need to know about the proportions of animals to make drawings. In this activity, students will find out about the proportions for cats' bodies—the relationship

between a cat's body length and its tail length. This information would be helpful to people who want to draw cats.

2. Have students make hypotheses (question 1).

   - What do you think the data will show about the relationship between a cat's tail length and its body length? Are cats' tails typically as long as their bodies? Half as long?

3. Students begin thinking about the relationship between tail length and body length by looking at an example cat (question 2). Beauportes has a tail 12 in. long and a body 28 in. long. Students can estimate that his tail is less than half his body length. The ratio can be described as $\frac{12}{28}$, $\frac{3}{7}$, 3:7, or 0.43.

   Ask students to explain how they figured out the relationship between the tail and the body length. This will bring up the need to divide the tail length by the body length. (*Note:* The tail lengths tend to be shorter than the body lengths, so dividing tail lengths by body lengths will result in ratios that are less than 1.)

4. Introduce the strategy of using TinkerPlots to make a formula to calculate the ratio of tail length to body length. This allows students to find the ratios for all the cats without having to do the calculations for each one. You may want to do a whole-class demonstration of the steps for making a formula using a projection system before students go through the steps on their own.

Exploration

5. Students create the formula themselves and then look for cats that have specific ratios (questions 4–11). The questions ask students to visualize what these cats look like. This can be difficult for some students because the ratios are decimals. Starting with the ratio of 0.50 can help students to visualize what a cat would look like with a tail length that is half the length of the body. The activity moves from looking for individual cats, to comparing pairs of cats, to analyzing the whole group of cats.

6. To summarize their findings, students write recommendations for drawing cats that will have tail lengths and body lengths in typical proportions (question 12).

Students need to add a text box to their plot and print it. To add the text box, drag a new text box from the shelf and drop it into the document. Before printing, have students choose **Show Page Breaks** from the **File** menu and move their plot so it appears on one page.

Wrap-Up

7. Discussion questions:
   - What is the relationship between the lengths of cats' tails and their bodies for this group of cats?
   - How did you analyze the data? What did you find out about this group of cats?
   - What are your recommendations for drawing typical cats? How long should you draw the tail in comparison to the body?
   - What range of *Tail_to_Body* ratios is reasonable? How long or short would a cat's tail have to be in comparison to its body for it to *not* look realistic?

ANSWERS

Tail Length and Body Length

1. Sample answer: From the pictures, I would expect most cats' tails to be somewhere between one-half and three-quarters as long as their bodies.
2. Beauportes has a tail 12 in. long and a body 28 in. long. Students can estimate that his tail is less than half his body length. The ratio can be described as $\frac{12}{28}$, $\frac{3}{7}$, 3:7, or 0.43. Methods will vary, but should include the idea of dividing tail length by body length.

8. There are 12 cats with a 0.50 ratio. Students need to find two.

Name of cat	*Tail_to_Body* ratio	*Tail_Length* (in.)	*Body_Length* (in.)
Co	0.50	**11**	**22**
Cleopatra	0.50	**9**	**18**
Wiley	0.50	**8**	**16**
Sinbad	0.50	**11**	**22**
Leo	0.50	**8**	**16**
K.C.	0.50	**12**	**24**
Chubbs	0.50	**11**	**22**
Tigger	0.50	**10**	**20**
Oscar	0.50	**11**	**22**
Lenny	0.50	**8**	**16**
Smudge	0.50	**10.5**	**21**
Hanna	0.50	**6**	**12**

9. Sample answer: Cats with a *Tail_to_Body* ratio of 0.50 have tails half as long as their bodies.

10. Sample answer: Gabriel's tail is almost $\frac{3}{4}$ the length of his body. In contrast, Boggie's tail is about $\frac{1}{3}$ the length of his body. Gabriel's body is about half as long as Boggie's, so even though their tails are the same length, the ratios of tail length to body length are very different.

11. Sample answer: Kiki and Flipturn have the same body lengths but different tail lengths. Flipturn has a longer tail, so his *Tail_to_Body* ratio is larger than Kiki's. Flipturn's ratio is a little larger than one half and Kiki's is a little smaller. I think the two cats would look similar.

12. Sample answer: I recommend drawing tail lengths that are about half as long as body lengths. There is a center clump of cats between 0.47 and 0.62, which means that 0.50 is a typical ratio. The mode is 0.50, which means that it was the most common ratio. Both the mean and the median for the *Tail_to_Body* ratio are close to 0.50, or $\frac{1}{2}$.

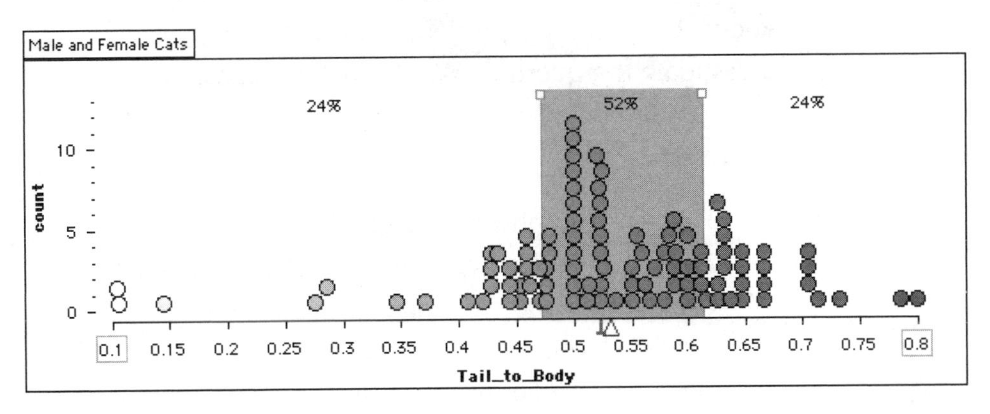

This graph shows that 69% of the cats had a *Tail_to_Body* ratio between 0.50 and 0.749 inclusive. This suggests that it makes sense to draw tails that are about $\frac{1}{2}$ to $\frac{3}{4}$ as long as the body lengths. However, our data show ratios of 0.11 to 0.80, and this is a small sample of cats, so it's realistic to expect tail lengths anywhere in this range and probably a few outside of it.

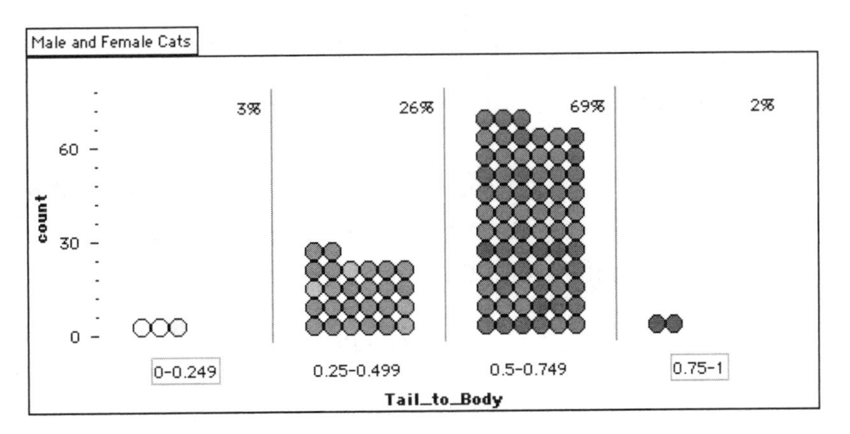

Tail Length and Body Length

Name:

ASK A QUESTION AND MAKE A HYPOTHESIS

If you want to draw pictures of cats, it is helpful to know about cats' proportions. Your task is to investigate the question: What is the relationship between the lengths of cats' tails and the lengths of their bodies?

Marc Henrie/Dorling Kindersley Collection © Getty Images

1. Here are two photos of cats. Use these pictures to help you make a hypothesis: What do you think the data for 100 cats will show about the relationship between a cat's tail length and its body length? For example, are cats' tails typically the same length as their bodies? Or are they longer or shorter?

Tierbild Okapia/Photo Researchers, Inc

ANALYZE DATA

2. First you'll look at a cat from the data set. Remember that the body lengths were measured from the head to the end of the body and do not include the tails.

   Beauportes: *Tail_Length* = 12 inches   *Body_Length* = 28 inches

   What is the relationship between tail length and body length for this cat? How did you figure this out?

3. Open the TinkerPlots file **Male and Female Cats.tp** to see a data set of 100 cats of different ages. You can use TinkerPlots to create a new attribute that will calculate the ratios of cats' tail lengths to their body lengths. Let's call the new attribute *Tail_to_Body.*

Follow these steps to make the formula for *Tail_to_Body.*

4. In the Attribute column of the data card, double-click **<new attribute>** and type `Tail_to_Body`.

5. Expand the stack of data cards until you see the Formula column, and double-click the circle for *Tail_to_Body.*

6. Enter the formula. You can type the names or expand the Attributes list and double-click the names to enter them.

7. Click **OK** when you are done. All the data cards will show the values for the attribute *Tail_to_Body.*

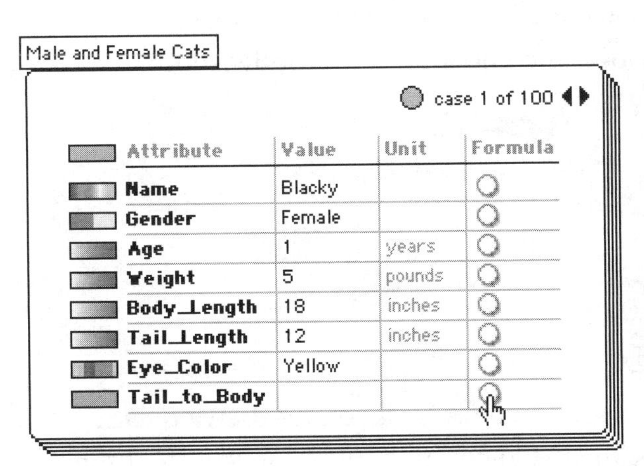

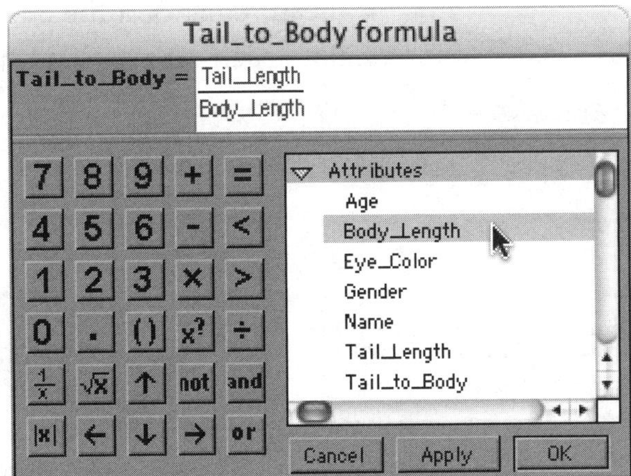

8. Find two cats that have a *Tail_to_Body* ratio of 0.50.

Name of cat	*Tail_to_Body* ratio	*Tail_Length* (in.)	*Body_Length* (in.)
Hanna	0.50	6	12
	0.50		
	0.50		

9. What do these cats look like? Write a description or draw a picture.

10. Use a line plot to find two cats that have tail lengths of 10 inches, but different *Tail_to_Body* ratios. Write the names and measurements of the two cats. How do the two cats compare? How does each cat's tail length look in comparison to its body length?

11. Use a line plot to find two cats with the same body lengths of 21 inches, but different *Tail_to_Body* ratios. Write the names and measurements of the two cats. How do the two cats compare? How does each cat's tail length look in comparison to its body length?

COMMUNICATE CONCLUSIONS

12. Now you will come up with your recommendation for drawing cats. How long should you draw the tail in comparison to the length of the body for a typical cat?

    a. Write your recommendation on a separate sheet of paper. You can draw a picture to help explain your recommendation.

    b. Choose a plot to convince your classmates that your recommendation makes sense. Add a text box to the plot to explain what it shows about the relationship between cats' tail lengths and body lengths. Print a copy of the plot.

Extending Ratios

OVERVIEW

Students create and analyze a new ratio to determine the relationship between cats' weights and their body lengths. This new ratio is more challenging than the one in Lesson 1.8; it is harder to visualize the relationship between weight and body length because they are measured in different units (pounds and inches). It tends to be easier for students to visualize the relationship between body length and tail length, because both attributes are measured in inches. Also, students are more familiar with lengths than weights. The lesson is designed to help students make sense of this new ratio, and to apply their understanding to analyzing the ratios for the group of cats.

The optional activity Is Feather Really Light? is designed to be similar to Lesson 1.3, which will give students the opportunity to show what they have learned about making comparisons and writing conclusions. This task is more challenging because it involves analyzing a larger data set (100 cats instead of 20). The task gives students the opportunity to use a variety of strategies, including determining what's typical by using dividers, percentages, and medians; exploring two attributes; and making a formula to create a new attribute. You can use this activity as an assessment of the section.

Objectives

- Use a formula to create a new attribute that will be helpful for making comparisons
- Use ratios to make comparisons
- Interpret and visualize the meaning of ratios: What does a cat look like with a specific weight to body length ratio? What does a low ratio mean in this context? A high ratio?
- Compare ratios
- Use a variety of data analysis methods to make comparisons

TinkerPlots **Class Time:** One or two class periods, depending on whether you do the optional activity.

Materials

- How Many Pounds per Inch? worksheet (one per student)
- Is Feather Really Light? worksheet (one per student, *optional assessment*)
- Data Analysis Vocabulary: Essential Data Analysis Terms worksheet (one per student, on CD)
- One stick of butter (*optional*)

Data Set: Male and Female Cats.tp (100 male and female cats of different ages)

TinkerPlots Prerequisites: Students should be familiar with intermediate graphing and creating new attributes by making a formula.

LESSON PLAN

Introduction

1. Introduce the missing cats story (question 2). Ask students to imagine that they are working in an animal shelter that has 100 cats, the ones in the data set. Have students investigate the data set with TinkerPlots to look for a cat or cats that might be Tracker.

2. Discuss students' strategies.

   - How did you figure out which cat or cats might be the missing one, Tracker?

   In the discussion, students will share how they determined whether or not a cat is skinny. This should bring up the need for analyzing the relationship between a cat's weight and its body length.

3. Introduce the ratio of weight to body length, which can be called the *Weight_to_Body* ratio and measured in units of pounds per inch (lb/in.). This ratio tends to be more difficult for students to visualize than *Tail_to_Body*. It helps to go through some examples.

   - If a cat weighs 12 lb and is 12 in. long, then the *Weight_to_Body* ratio is 1 lb/in.
   - If a cat weighs 6 lb and is 12 in. long, then the *Weight_to_Body* ratio is 0.50 lb/in. Another way to say this would be $\frac{1}{2}$ lb/in.

For most of the cats, the values for weight are smaller than the values for body length, so their *Weight_to_Body* ratios are less than 1.

You can also use a stick of butter as a visual example of a weight to length ratio. Typically, a stick of butter weighs 4 oz and is about 4.5 in. long. Ask students: How much does each inch of butter weigh? The ratio of weight to length for the butter is 0.89 oz/in., or about $\frac{9}{10}$ oz/in.

Exploration

4. Students use TinkerPlots to make a formula for *Weight_to_Body* ratio (question 4). They can apply their experience making a formula in Lesson 1.8 to the task.
5. Students find different cats with specific ratios, which will help them build understanding of the meaning of the ratios (question 5). Then they compare the group of male cats to the group of female cats (question 7).
6. Discuss the class results.
   - Which cat did you choose for each description? Why?
   - Do you think the cat in question 7 is more likely to be male or female? Why?
7. If you are using the optional activity Is Feather Really Light?, ask students to think back to Lesson 1.3, when they compared Chubbs to a group of 20 one-year-old cats. Introduce the new question: Is the cat named Feather light in comparison to the group of 100 cats?
8. Students work independently to compare Feather to the group of 100 cats. They select a plot that they think is particularly helpful for making the comparisons, and write their conclusions.
9. Have students share their conclusions with the class.
   - Which attributes did you compare? Why?
   - What kinds of plots did you make? Why?
   - What are your conclusions?
10. Have students compare these conclusions with the ones that they wrote for Lesson 1.3 about the cat named Chubbs.

Wrap-Up

11. Discussion questions:

    - What have you learned about analyzing data and writing conclusions since the start of this section?

    - What are some different ways to compare groups?

12. To summarize some of the concepts in this section, ask students to complete the Data Analysis Vocabulary: Essential Data Analysis Terms worksheet from the *Digging into Data* CD. Have students share their ideas with the class.

ANSWERS

How Many Pounds per Inch?

2. Sample answer: I think Tracker is the cat Boggie, case 62. We know Tracker's weight is 10 lb, so I made a line plot of weight to find all the 10-lb cats. Then, I thought a cat described as skinny should have a long body, so I graphed body length on the vertical axis. I colored by *Gender* to look for the longest male at 10 lb, which is Boggie.

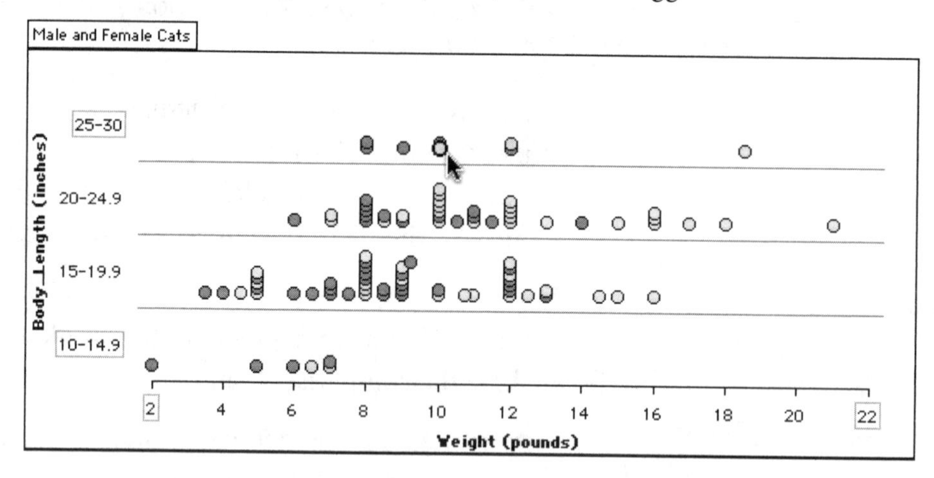

3. a. Harmony: 0.50 lb/in.; Scooter: 0.80 lb/in.

   b. Sample answer: I divided the weight by the body length.

5. There are eight cats with a *Weight_to_Body* ratio of 0.50 lb/in. and two cats with a ratio of 0.25 lb/in. Students need to find only one of each.

Ratio	Name of cat	*Weight_to_Body* ratio (lb/in.)	*Weight* (lb)	*Body_ Length* (in.)
Lowest ratio	**Fuzzy**	**0.15**	2	13
Highest ratio	**Domino**	**1.05**	21	20
0.50	**Fluffy**	0.50	10	20
0.50	**Leo**	0.50	8	16
0.50	**Gabriel**	0.50	7	14
0.50	**Lenny**	0.50	8	16
0.50	**Harmony**	0.50	12	24
0.50	**Zero**	0.50	12	24
0.50	**Rascal**	0.50	10.5	21
0.50	**Tabby Burton**	0.50	10	20
0.25	**Pywacket**	0.25	4	16
0.25	**Sparky Briles**	0.25	4.5	18

6. Sample answer: A cat with a *Weight_to_Body* ratio greater than 1 lb/in. would weigh more than one pound for every inch of its body length. Almost all the ratios are below 1 lb/in., so this would be a very fat cat!

7. Sample answer: The cat in the cartoon is probably male. He is pretty heavy, and looks like he has a high weight to body length ratio. In our data set, males tend to have higher ratios (0.43 to 0.70 lb/in.) than females (0.35 to 0.53 lb/in.), and males have all the ratios above 0.80 lb/in.

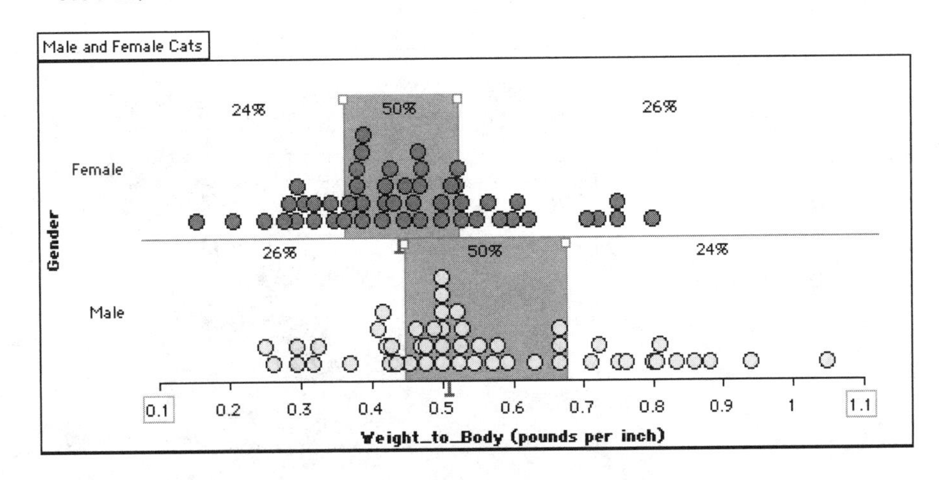

Is Feather Really Light?

2. Feather is case 74. He weighs 13 lb.

3.–5. Feather is not light compared to this group of cats. Here are some characteristics of a strong response.

- Student draws correct conclusion: Feather is not light.
- Student examines both the attributes of weight and body length, or the ratio of weight to body length.
- Student creates a plot that clearly compares Feather with the group of cats, and explains what this plot shows about Feather.
- Student writes strong and specific statements to support conclusions. Statements match criteria for high-quality conclusions from the rubric for making comparisons.

Sample plots and analysis: These two plots show that 85% of the cats weigh less than Feather. In addition, Feather's weight is above the median and mean weights for the cats.

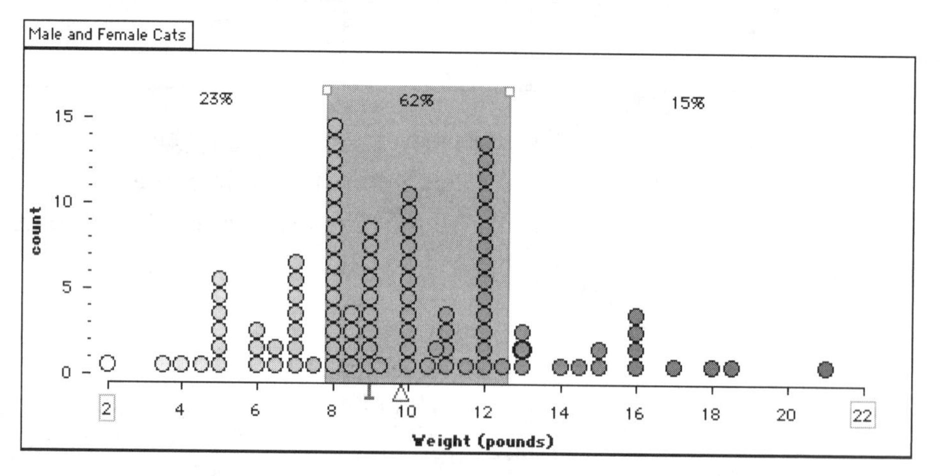

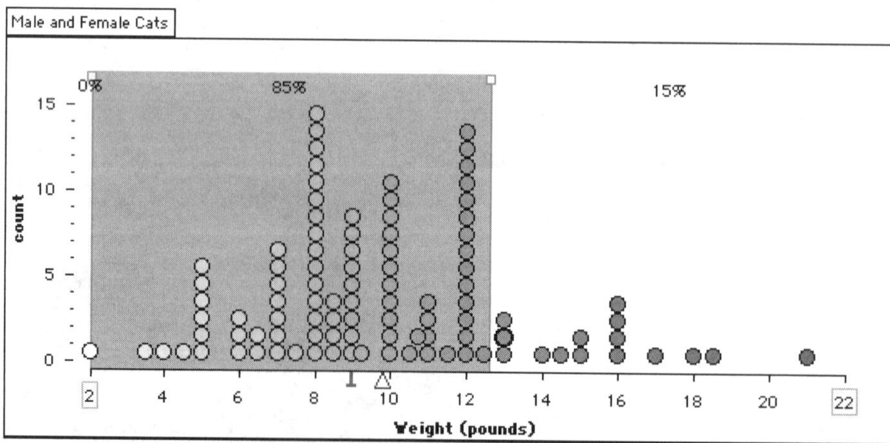

Feather is slightly below the mean and median for body length, which suggests that he has a medium body length. He falls in the center clump of cats for body length, which shows that he has a typical body length. The combination of a high weight and a medium body length shows that Feather is not a light cat.

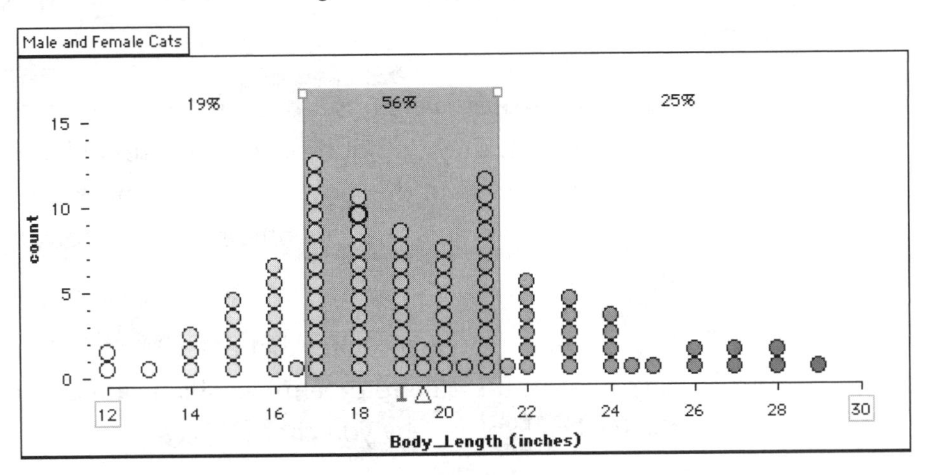

Feather has a weight to body length ratio of 0.72 lb/in. This is in the top 25% for the group, which means that he is one of the heavier cats.

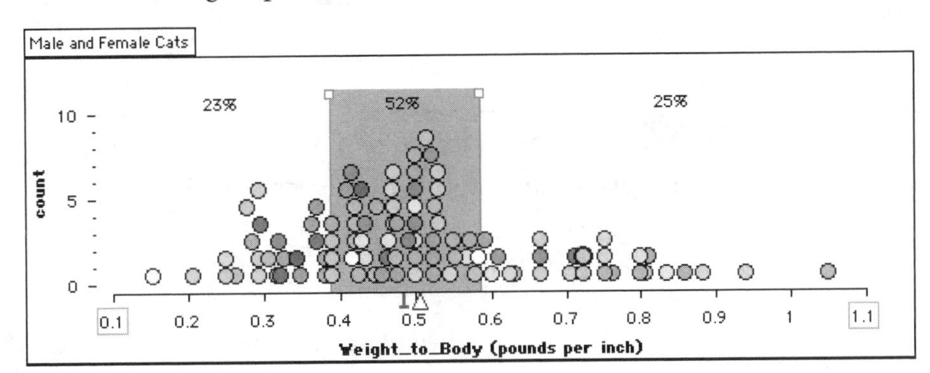

How Many Pounds per Inch?

Name:

In this lesson, you will look at the ratio of a cat's weight to its body length.

1. Open the TinkerPlots file **Male and Female Cats.tp** to see data from 100 cats.

2. Sara and Mark are very upset because their cat, Tracker, is missing. Unfortunately, they do not have any pictures of him. They describe Tracker as friendly and skinny. At the cat's checkup last week, he weighed 10 pounds. They are offering a $200 reward for the return of Tracker.

   You work at an animal shelter that has 100 cats (the ones in the data set). Analyze the data to determine which cat best fits the description for Tracker. Explain why you chose this cat.

3. Your next task is to create a new attribute that will calculate the ratio of cats' weights to their body lengths. This ratio will tell you how many pounds a cat weighs per inch of body length. Let's call the new attribute *Weight_to_Body.*

   a. What are the *Weight_to_Body* ratios for these two cats from the data set?

      Harmony: *Weight* = 12 pounds, *Body_Length* = 24 inches,
      Weight_to_Body ratio = ________ pounds per inch

      Scooter: *Weight* = 16 pounds, *Body_Length* = 20 inches,
      Weight_to_Body ratio = ________ pounds per inch

   b. How did you figure out the ratios?

4. Use TinkerPlots to create a formula to calculate this ratio for all the cats in the data set.

5. Find one cat with each *Weight_to_Body* ratio. Round answers to the hundredths.

Ratio	Name of cat	*Weight_to_Body* ratio (pounds per inch)	Weight (pounds)	*Body_Length* (inches)
Lowest ratio				
Highest ratio				
0.50		0.50		
0.25		0.25		

6. What does it mean if a cat has a *Weight_to_Body* ratio that is greater than 1 pound per inch? What does this cat look like?

7. Do you think that the heavy cat in the cartoon is more likely to be male or female? Make sure that you use the data to back up your answer.

Is Feather Really Light?

Name:

ASK A QUESTION

In a previous lesson, you compared the cat Chubbs to a group of 20 cats. Now, your task is to analyze the data for 100 cats to answer the question: Is the cat named Feather light compared to this group of cats?

1. Open the TinkerPlots file **Male and Female Cats.tp.**

2. Make plots to find the cat named Feather by using the information given. Fill in the blanks.

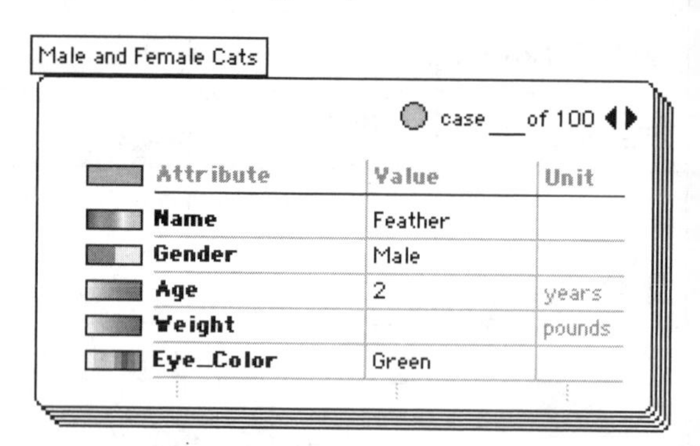

ANALYZE DATA

3. What attributes will you analyze to compare Feather to the group of cats? You may want to use formulas in TinkerPlots to create new attributes.

4. Analyze the data by creating different plots. Choose one plot that you think is particularly helpful for comparing Feather to the group of cats.

   a. Highlight the dot for Feather so that other people can find him.

   b. Add a text box to the plot to explain what the graph shows about Feather in comparison to the other cats.

   c. Print a copy of the plot.

COMMUNICATE CONCLUSIONS

5. Write your conclusions on a separate sheet of paper for the question: Is Feather light in comparison to the group of cats? Make sure to use evidence from the data to support your conclusions.

Section 2

Comparisons and Box Plots: Investigating Data about Middle-School Students

In this section, students investigate questions by comparing data for different attributes and groups. They focus on making visual comparisons: comparing the overall shapes of distributions and the positions of center clumps. They also make numeric comparisons by using the median, mean, and range. The lessons help students move from using line plots to using box plots, a more abstract representation. The data sets are taken from surveys of sixth-, seventh-, and eighth-graders about topics that students find easy to identify with. You have the option of conducting your own class survey and having students analyze that data, as well.

OBJECTIVES

Asking Questions about Data

- Pose questions that can be investigated about a data set (Lesson 2.4)
- Come up with hypotheses (Lessons 2.3, 2.4, 2.7)

Collecting Data

- Write survey questions (Lesson 2.1)
- Conduct a survey to collect data (Lesson 2.1)

Analyzing Data

- Reason about what the shapes of distributions might be for different attributes (Lesson 2.2)
- Make visual comparisons by analyzing the shapes of distributions (Lesson 2.3)
- Make numeric comparisons by analyzing means, medians, and percentages (Lesson 2.3)

- Deepen understanding of the term *typical* and of how to determine what's typical for a group (Lessons 2.3, 2.6, 2.7)
- Compare two or more attributes for the same group (Lesson 2.3)
- Compare two or more groups (Lessons 2.6, 2.7)
- Experiment with different ways to display data, including dividing the data into equal intervals (Lessons 2.2, 2.4)
- Create, interpret, and compare line plots (Lessons 2.2, 2.3, 2.5)
- Build understanding of the measures in the five-number summary, and how these are used to create box plots (Lessons 2.5, 2.6)
- Learn the terms *quartile, interquartile range, maximum,* and *minimum* (Lessons 2.5–2.7)
- Create, interpret, and compare box plots (Lessons 2.5–2.7)
- Compare two representations of data: line plots and box plots (Lesson 2.5)

Communicating about Data

- Describe and summarize data (Lesson 2.2)
- Communicate conclusions orally and in writing (Lessons 2.3–2.7)
- Write comparison statements that use the data as evidence (Lesson 2.7)

Applications of Math Concepts for Other Strands

- Number and Operations: Use percentages to analyze and compare data sets (Lessons 2.3, 2.5–2.7)

TINKERPLOTS SKILLS

This section builds on the TinkerPlots skills introduced in Section 1: basic graphing; creating line plots, using dividers, and showing percentages, means, and medians. Students learn new TinkerPlots skills, including duplicating a plot, changing axis endpoints, using a color key, showing box plots, and hiding icons. If your students did not do Section 1, you can introduce the TinkerPlots skills needed for Section 2 through class demonstrations or by having students watch the *TinkerPlots Basics* movie (available from the **Help** menu).

Introducing Survey Data

OVERVIEW

The main priority of this lesson is to introduce students to the context and attributes of the data sets that they will be working with in this section. Students answer survey questions and write their own questions to help them become familiar with using surveys to collect data. This is a shift from Section 1, in which students worked with data collected primarily by taking measurements. The lesson provides the option for preparing and conducting a survey, so that students will be able to investigate their own data in Lesson 2.4. This process can be extremely valuable and motivating for students. However, the data collection process can be time-consuming, so this is an optional rather than essential component of the section.

Objectives

- Learn about the data set
- Become familiar with the survey used to collect the data
- Generate different types of survey questions that can be used to collect quantitative and categorical data
- Build understanding of the characteristics of high-quality survey questions
- Collect data by using a survey

Offline

Class Time: One class period. *Note:* Additional time is needed for the optional survey activity—to prepare and make copies of the surveys, to conduct the surveys, and to enter the data into TinkerPlots.

Materials

- Seventh-Grade Surveys worksheet (one per student)
- Survey worksheet (one per student)
- Calculator (one per student, *optional*)

LESSON PLAN

Parts 1–4 of this lesson introduce the data set and the survey questions. Parts 6–8, on preparing and conducting a class survey, are optional.

Introduction

1. Introduce the context of the data set: 64 seventh-graders took a survey. Have students take the survey themselves to get familiar with the questions and the attributes (question 1). For the second part of the survey, have students determine the amount of time they spend on different activities during a typical week. Encourage students to make realistic estimates and to check their calculations. You may want to have students use calculators for this. Students should check their estimates against what is plausible—for example, they can't talk on the phone for more than 24 hours a day.

2. Have students individually read the two sample data cards and write questions about the data (questions 2 and 3). Discuss.

   - For which four attributes did the two students respond the same way?
   - What questions did you come up with about the information on the data cards?

   Note: Lesson 1.1 introduces the definition of *attribute.* If your students did not do Lesson 1.1, then go over the definition here. An *attribute* is a characteristic of a person or thing. *Variable* is another term for attribute.

Exploration

3. Discuss the qualities of a well-written survey question. Have the class brainstorm a list of qualities, such as "clearly written," "easy to understand," and "not ambiguous." If a question is confusing, people may interpret it in different ways, and it will be hard to make sense of their answers. For example, if you ask, "How much time do you spend reading?" it's not clear if you mean during a typical day, typical week, or specific day, or if people should give their responses in hours or minutes. It's also important that the survey questions respect people's privacy and are not hurtful or too personal. Discussion questions:

   - What might be confusing about this survey question: How much time do you spend reading? How would you change the question to make it clearer?
   - Look at the list of survey questions. Did you find any of the questions confusing? How would you change them to make them clearer?
   - What are the characteristics of a high-quality survey question?

4. Have students work in small groups to brainstorm survey questions that they would like to ask students in their school. Ask them to write both questions that can be answered with a number and questions that can be answered with a word (questions 3 and 4).

Wrap-Up

5. Have students exchange their survey questions with other students and give each other feedback to improve the questions.

Prepare and Conduct Your Own Survey (optional)

There are many benefits to preparing and conducting a class survey because students tend to be very invested in and motivated to investigate their own data. To give you time to collect the data and enter it into TinkerPlots, the next three lessons use prepared data sets, and then Lesson 2.4 uses your data. If you need more time to prepare the data, you can use Lesson 2.4 at any time later in the unit.

6. You have several options for preparing a survey to collect your own data. You can use the Survey worksheet, perhaps revising it by having your students add new questions or by cutting some questions. Alternatively, you can start from scratch and have your students create a new survey. Here are some suggestions for creating your own survey.
   - Use a mix of quantitative and categorical questions. It's helpful to have more quantitative questions because these lead to richer investigations.
   - Use about 6–12 questions to allow enough variety of attributes for students to investigate.
   - Keep the survey anonymous. If respondents include their names, then when students analyze the data they may focus more on individuals than on thinking about the group of students (which is central to the math goals).
   - Format the survey so that there is a column on the right side for students to write their answers (see the Survey worksheet). This will make it easier to transfer the information to TinkerPlots.

- Collect data from a sample of 50–100 students, so that students will have a rich data set to investigate.
- If possible, administer the survey to two different groups, such as two different grade levels, so that students will be able to make comparisons.

7. Decide how you will administer the survey. If you are collecting data only within the class, then you can have each student fill out a survey during class time. You could repeat this process, using the same survey with different class sections. If the class is going to collect data from other students in the school or from adults, then discuss a plan for conducting the survey. Who will they ask? When? How many people? Discuss the importance of administering the survey in a way that respects respondents' privacy.
8. Once you have the data, there are several ways to get it into a TinkerPlots data set. You might want to watch the TinkerPlots movie *Adding Data* for more information about these methods. (Choose **TinkerPlots Movies** from the **Help** menu.)
   - Enter the data in a case table in TinkerPlots. In a blank document, drag a new table into the document. Click <**new**>**,** type an attribute name, and press Enter. Type the values in the column. (In a case table, each row corresponds to one case.) The data cards will be filled in automatically.
   - Set up a TinkerPlots data set by typing the attributes names on a data card. Have students take turns entering their data on a card in the same file on the same computer. When all the students have entered their data, copy the file onto the other computers.
   - Enter the data into an Excel spreadsheet, then copy the data into TinkerPlots. (See TinkerPlots Help for details.)

ANSWERS

Seventh-Grade Surveys

1. Survey answers will vary.
2. a. *Age, Household, Countries_Lived, Letters_Name*

b. Sample questions:

How much more time does the female student spend getting to school than the male student? [9 min]

How does each student get to school? [The male student walks to school, and the female student takes the bus.]

Which student has lived in more states? [The female student has lived in more states.]

Which student has more pets? [The female student has more pets.]

3. Sample questions:

Survey question	Attribute name; unit	Range of values
How much time do you spend listening to music in a typical week?	*Music_Time;* unit: hours	0–30 h
How much time do you spend practicing a musical instrument in a typical week?	*Practice_Time;* unit: hours	0–15 h
How many brothers do you have?	*Brothers*	0–8
How many times a week do you eat a sandwich for lunch?	*Lunch_Sandwich*	0–7

4. Sample questions:

Survey question	Attribute name	Responses
What is your favorite kind of music?	*Music_Type*	Rock, Rap, Jazz, Classical, Metal, Other Most common: **Rock**
What is your favorite type of book to read?	*Book_Type*	Fiction, Science fiction, Mystery, Biography, Nonfiction, Other Most common: **Fiction**

Seventh-Grade Surveys

Name:

Surveys are useful tools for collecting data. You will take a survey, look at data from seventh-graders, and write your own survey questions.

1. Read and fill out the survey. This survey was filled out by 64 seventh-graders. You will be analyzing their data.

2. These data cards show two students' responses to the first 11 survey questions.

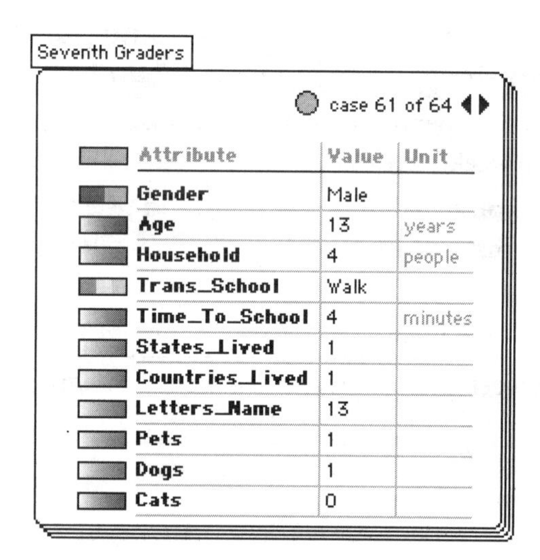

Seventh Graders

case 61 of 64

Attribute	Value	Unit
Gender	Male	
Age	13	years
Household	4	people
Trans_School	Walk	
Time_To_School	4	minutes
States_Lived	1	
Countries_Lived	1	
Letters_Name	13	
Pets	1	
Dogs	1	
Cats	0	

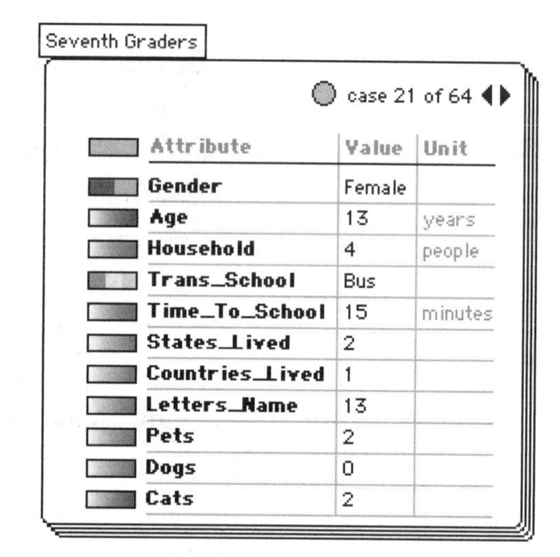

Seventh Graders

case 21 of 64

Attribute	Value	Unit
Gender	Female	
Age	13	years
Household	4	people
Trans_School	Bus	
Time_To_School	15	minutes
States_Lived	2	
Countries_Lived	1	
Letters_Name	13	
Pets	2	
Dogs	0	
Cats	2	

a. These two students have the same responses for four attributes. Circle those attributes on the cards.

b. Write two questions that can be answered from the information on the data cards. Then, write the answers for your questions.

Question	Answer
Which student spends more time getting to school?	The female student spends more time getting to school.

3. What survey questions would you like to ask students in your school? For each question, give the units, if any, and the attribute name for the data card. Also tell what you think the range of values would be if you asked 100 students. Write two questions that students could answer with a number.

Survey question	Attribute name; unit	Range of values
How many hours do you sleep on a typical school night? ________ hours	Hours_Sleep; unit: hours	4–12 hours
a.		
b.		

4. Write two survey questions that students would answer using a word or a letter. Give at least three responses for people to choose from. What do you think the most common response would be?

Survey question	Attribute name	Responses
How do you get to school on a typical day?	Trans_School	Walk, Car, Bus, Other Most common: Car
a.		
b.		

Survey

Name:

Attribute name	Survey question	Your answer
1. *Gender*	What is your gender?	Male Female
2. *Age*	How old are you?	________ years
3. *Household*	How many people live in your household, including yourself?	
4. *Trans_School*	How do you get to school on a typical day?	Walk Car Bus Other
5. *Time_To_School*	How long does it take you to get to school?	________ minutes
6. *States_Lived*	How many states have you lived in?	
7. *Countries_Lived*	How many countries have you lived in?	
8. *Letters_Name*	How many letters are there in your full name (first and last names, but not middle)?	
9. *Pets*	How many pets do you have?	
10. *Dogs*	How many dogs do you have?	
11. *Cats*	How many cats do you have?	

How many hours do you spend on different activities in a typical week? Use the table to estimate. Round your answers to the nearest half hour.

Attribute name	Activity	School day (h)	Weekend day (h)	Typical week (h)
12. *Sports_Time*	Playing sports or exercising			
13. *Homework_Time*	Doing homework			
14. *Phone_Time*	Talking on the phone			
15. *Music_Time*	Listening to music			

OVERVIEW

This lesson immerses students in the seventh-graders data set and helps them get to know the different attributes. Students build their understanding of distributions by visualizing what the distribution of data might look like for particular attributes. They work with two representations: intervals and line plots. This builds on their work in Section 1 and helps prepare them for Lesson 2.3.

Objectives

- Experiment with different ways to display data with TinkerPlots
- Deepen understanding about distributions by reasoning about what the distribution of data for different attributes might look like
- Create and interpret line plots and interval plots
- Write summary statements about a graph

TinkerPlots

Class Time: One class period

Materials

- Name the Mystery Attributes worksheet (one per student)

Data Set: **Seventh Graders.tp** (survey data from 64 seventh-graders)

TinkerPlots Prerequisites: Student should be familiar with basic graphing, showing counts and percentages, and using the **Drawing** tool.

LESSON PLAN

Introduction

1. Introduce the task of figuring out what the mystery attributes are and creating the graphs with TinkerPlots. If students have done Lesson 1.2, point out that this activity is similar, but is more challenging because it involves a larger data set and more complex graphs. Remind them to scroll down to see the different graphs and to move the data card as needed.

Exploration

2. Have students work independently or in pairs to make the mystery graphs (questions 2–7). This will help them to expand their repertoires of plots to make with TinkerPlots and prepare them for subsequent

lessons. If students are having difficulty figuring out what a mystery attribute is, you may want to use these questions.

- Which attributes might have this range of data?
- Are there any zeroes on the graph? Which attributes would have zeroes (or would not have zeroes)?
- What can you find out from looking at the shape of the data? Which attributes might look like that?

3. When students have finished, have a class discussion about the strategies they used.
   - How did you figure out what the mystery attributes were for each graph?
   - What questions can you ask yourself to get started?
   - What clues can you get from the shapes of the distributions?
   - What clues can you get from the range of the data?

Wrap-Up

4. Have students select a plot and write a summary about the data (question 9). Emphasize that they need to summarize what the graph shows about the *group* of seventh-graders. Ask a few students to share their summaries with the class.
   - What did you find out about the group of seventh-graders from your graph?
   - What does the shape of the distribution tell you about the group of students?

ANSWERS

Name the Mystery Attributes

2. *Household*
3. *Trans_School*
4. *Letters_Name*
5. *Cats*
6. *E: Pets; F: Gender*

7. *G: Time_To_School; H: Trans_School*

8. Sample student work:

   Kiara: I figured out that mystery attribute B was TransSchool by looking at the graph. The data was broken into 3 parts in [and] there are 3 types of transport: car, bus, walk.

   Comments: The student figured out that there were three types of transportation and that the graph was in three parts.

   D.J.: I figured out that mystery attribute D was cats because most people don't have a cat so the mode would be 0.

   Comments: The student used the zeroes as a clue for figuring out which attribute was graphed.

   Eana: I figured out that mystery attribute A was household by looking at the graph and seeing which one makes the most sense and then trying it.

   Comments: The student's answer is too general. She needs to explain how she figured out that the graph represented data for the attribute *Household.*

9. Sample student work:

   Kiara: Graph Title: Letters in 7th Graders' names. Summary Statements: The data is bimodal (10, 12) and the range is 12. There is a large clump between 9 and 14. The data is skewed to the left.

   Comments: The student is using a variety of data analysis terms but does not explain what the terms mean in the context of the data about students' names. She is using the term "skewed to the left" informally to describe that the clump is on the left side of the scale. Technically the data are skewed right because the tail of the distribution extends in that direction.

Name the Mystery Attributes

Name:

1. Open the TinkerPlots file **Seventh Graders.tp.** This data set is from 64 seventh-graders who filled out a survey.

Each graph shows one or two attributes from the Seventh Graders data set. For each graph, figure out what the mystery attribute is and then make the graph with TinkerPlots. (Scroll down to see the different graphs. If you need to move the data card, select it and drag the top border.)

2. What is mystery attribute *A*? Make the graph.

   A: ________

3. What is mystery attribute *B*? Make the graph.

   B: ________

4. What is mystery attribute *C*? Make the graph.

   C: ________

5. What is mystery attribute *D*? Make the graph.

   D: ________

6. What are the mystery attributes? Make the graph.

   E: ________ *F*: ________

7. What are the mystery attributes? Make the graph.

   G: ________ *H*: ________

Answer these questions on your own paper.

8. How did you figure out what the mystery attributes were? Explain the strategy you used for one of the graphs.
9. Choose one graph and analyze the data. Come up with a title for the graph. Write at least two statements to summarize what you found out from the graph about the attribute or attributes.

Comparing Two Attributes

OVERVIEW

In this lesson, students compare two attributes (*Sports_Time* and *Phone_Time*) for the same group of middle-school students. Students need to create two graphs that have the same scale to make this comparison. To determine what's typical for each attribute, students use dividers to find the middle 50% of the values on each line plot. This approach to dividing the data into three parts (25%, 50%, 25%) will help prepare students for working with box plots in Lessons 2.5–2.7.

Objectives

- Make visual comparisons by analyzing the shapes of distributions
- Determine what's typical for a group by finding the middle 50% of the values
- Make numeric comparisons by examining means, medians, and percentages
- Understand the importance of scale when comparing two graphs
- Communicate conclusions using the data as evidence

TinkerPlots

Class Time: One class period

Materials

- Sports Time and Phone Time worksheet (one per student)
- Rubric for Making Comparisons (*optional,* on CD)

Data Set: Students Time.tp (data on 122 middle-school students)

TinkerPlots Prerequisites: Students should be familiar with basic graphing; showing mean, median, and percentages; and using the **Drawing** tool and dividers.

TinkerPlots Skills: Duplicating a plot, changing endpoints, and using the color key are explained in this lesson.

LESSON PLAN

Introduction

1. Ask students to write their hypotheses for the question: Do middle-school students tend to spend more time talking on the phone or playing sports in a typical week?

2. Explain that the data were collected by students making estimates about the time they spent talking on a phone and playing sports in a typical week. If your students completed the survey in Lesson 2.1, then they will have made their own estimates about the amount of time they spend on different activities. Discussion questions:
   - What are some of the benefits and drawbacks of collecting data by having people make estimates about their time? [Benefits: easier than keeping logs of time; drawbacks: may be unreliable, students could exaggerate or underestimate, students could make math errors in calculating a typical week.]
   - What other data collection methods would you suggest using to find out how much time students spend on different activities?

Exploration

3. Have students work individually or in pairs to analyze the data. To help students make fair comparisons, the worksheet provides directions for displaying two plots that are the same size and have the same scale. If the two graphs have different scales or are different sizes, then students may get misleading impressions of how the distributions compare.

4. Students need to determine what's typical by applying what they learned in Lessons 1.5 and 1.6 about using dividers to find the middle 50% of the data.

   For question 8, show students how to use the **Drawing** tool to sketch the overall shape of the distribution. Point out that the focus is on the general shape, so students should not draw lines that go up and down with every bump in the data.

If your students did not do Lessons 1.5 and 1.6, then demonstrate how to do questions 8 and 9.

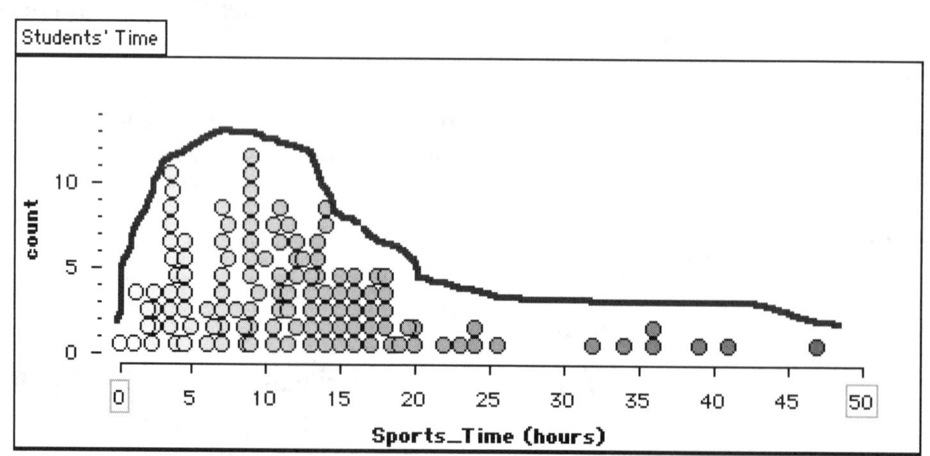

Ask students to compare the shapes of the two distributions. You may want to have the class brainstorm a list of words for describing the shapes, or use the word bank in Lesson 1.4.

For question 9, show students how to use the dividers to determine typical times for sports and phone. Demonstrate how to use the percentages to position the dividers so that they enclose the middle 50% of the values for *Sports_Time*. Then have students position the dividers themselves for *Phone_Time.*

5. Have a class discussion about using the dividers.

   - Why is this method of finding 50% of the values helpful for figuring out what's typical? [Finding 50% of the values is an impartial method of choosing a center clump—the percentages are not affected by the size of the graphs or the dots. Depending on how large or small you make the graph or the dots, where the dots cluster can look very different.]
   - Why is it important to find the middle 50% of the data, and not just any 50%? [We are trying to find what is typical of the group as a whole, so we are looking for the "center" of the distribution. The values in the center are more representative of what's typical for the group than the values at the extremes.]
   - Why isn't the "middle 50%" always positioned in the middle of the axis? (For example, the "middle 50%" for sports and phone time were positioned on the left side of the axes, not in the middle.) What is it the middle of? [It is the middle half of the values in the ordered data set. These values might not be in the middle of the range of the data if, for example, there are some very high values but no low values.]

Wrap-Up

6. Have students write their conclusions. If you are using the Rubric for Making Comparisons, go over it with the class. This rubric is similar to the one used in Lesson 1.7. *Note:* If your students did not do Lesson 1.7, you may want to spend an additional class period (offline) to introduce the rubric. Give examples of the different criteria. Have students write their conclusions, then work with a partner to improve them.

7. Have students share their conclusions with the class.

   - For this investigation, you had to make two graphs with the same scale. Why is it important for the graphs to have the same scale?

- What did you find out from comparing the shapes of the distributions?
- What were typical times for each activity? How did the two activities compare?

8. Have students consider how their findings might be different for other activities and other samples of people.
   - What do you think your findings would be if you compared other activities, such as TV time and Internet time?
   - What do you think your findings would be if you analyzed data for a different sample of people, such as adults?

ANSWERS

Sports Time and Phone Time

1. Sample answer: Students probably spend more time on sports.

8. Sample answers:

Similarities and differences in the shapes of the two distributions	What does this tell you about how the amount of time students spend on the two activities compares.
Both have more values near zero and have a few high values.	A few spend a lot of time on sports or the phone. Most spend fewer than 3 h on the phone and fewer than 13 h on sports.
The middle 50% of *Sports_Time* is higher on the scale and wider.	Students tend to spend more time on sports than on the phone.

10. Sample answer: Students typically spend about 6–15 h on sports, but only about 1–8 h on the phone.

11. Sample answers:

    Mean: *Sports_Time:* 12 h; *Phone_Time:* 7 h; Students tend to spend about 5 h more on sports.

    Median: *Sports_Time:* 11 h; *Phone_Time:* 3.5 h; Most students spend more time on sports (about 7.5 h more).

12. Conclusions should incorporate students' answers to questions 8–11. You might use the Rubric for Making Comparisons to evaluate them.

Sports Time and Phone Time

Name:

In this activity, you will compare two attributes.

ASK A QUESTION AND MAKE A HYPOTHESIS

Does this group of middle-school students tend to spend more time playing sports or talking on the phone during a typical week?

1. What do you think the data will show? Why?

ANALYZE DATA

In other lessons, you were able to compare two attributes on one plot. Now you need to create two plots: one for each attribute.

2. Open the TinkerPlots file **Students Time.tp.** 122 students took a survey in which they estimated their time on different activities during a typical week.

3. Make a line plot for the attribute *Sports_Time.*

4. Select the plot, and then go to the **Edit** menu and choose **Duplicate Plot.**

5. Click the **Key** button for the first *Sports_Time* plot. Click the lock to close it so that the colors will not change.

6. Drag the attribute *Phone_Time* to the horizontal axis of the second plot.

7. On the *Phone_Time* plot, double-click the box at the right end of the axis, and change the endpoint to 50 so both graphs have the same scale. This will make it easier to fairly compare the graphs.

Compare the Shapes of the Distributions

8. Use the **Drawing** tool to draw the overall shape of each plot. Compare the shapes of the distributions.

Similarities and differences in the shapes of the two distributions	What does this tell you about how the amount of time students spend on the two activities compares?

Compare Typical Times

9. Use the dividers and percentages to find the middle 50% of times for each graph. This example shows that 50% of the students spend about 6 to 15 hours per week playing sports.

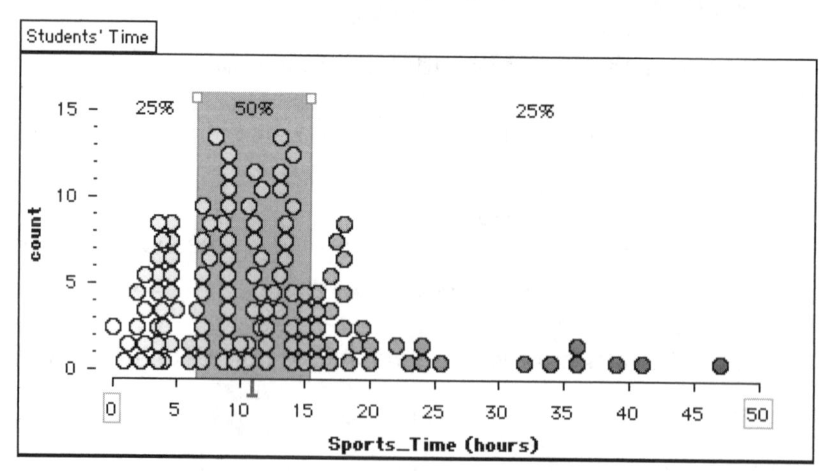

10. Compare the amount of time students typically spend doing sports to the amount of time students typically spend on the phone.

Sports: __________ to __________ hours per week

Phone: __________ to __________ hours per week

Make Numeric Comparisons

11. Use means and medians to compare the amount of time students spend on sports to the time spent talking on the phone.

Measure	***Sports_Time* (h)**	***Phone_Time* (h)**	**How do the measures compare? What does this tell you about how the students spend their time?**
Mean △			
Median ⊥			

COMMUNICATE CONCLUSIONS

12. What is your conclusion for the question: Does this group of middle-school students tend to spend more time talking on the phone or playing sports on a typical week? Use evidence from the data to support your conclusion.

Analyzing Class Survey Data

OVERVIEW

Students tend to be highly motivated to investigate their own data. This lesson gives students the opportunity to ask their own questions about the data that they collected in Lesson 2.1. They need to figure out how to analyze the data to answer their questions, and then write their conclusions. This is the first lesson in which students go through all the parts of a statistical investigation: asking a question, making a hypothesis, collecting data, analyzing the data, and writing a conclusion. In previous lessons, parts of the process were already done for them.

Objectives

- Ask questions about data that the class collected
- Figure out how to analyze the data to answer a question
- Apply a variety of data analysis methods to investigate the data set
- Communicate conclusions that use the data as evidence

TinkerPlots

Class Time: One class period

Materials

- Investigating Data worksheet (one per student, on CD in the file **Templates.pdf**)
- Copies of the survey used to collect the data (*optional*)

Data Set: Use a data set of survey data collected by your classes. See Lesson 2.1 for suggestions on preparing your class's data.

TinkerPlots Prerequisites: Students should be familiar with intermediate graphing.

LESSON PLAN

Introduction

1. Give students a few minutes to use TinkerPlots to get familiar with the data set and the attribute names. You may want to hand out copies of the survey to help students connect the attributes with the survey questions (especially if it's been a while since they took the survey).

   Note: Students often try to find themselves in the data set. This can be a fun way to get to know the data set. However, some students want to identify other students, which runs counter to having an anonymous survey and respecting the privacy of the respondents. Remind students that the focus is on analyzing the group as a whole, not on individuals.

2. Have students brainstorm questions that they would like to investigate in the class data (question 1). Encourage them to come up with questions that focus on the group of students and involve making comparisons by looking at more than one attribute. Examples:

   - How does the amount of time these students spend watching TV compare with the amount of time they spend listening to the radio?
   - How does the amount of time this group of females spend on the Internet compare with this group of males?

   You may also want to give some "non-examples" of questions that focus on individual cases instead of the group, such as: Who spends the longest time getting to school?

3. Have the class brainstorm a list of ways to analyze the data with TinkerPlots.

Exploration

4. Students use TinkerPlots to investigate their own questions about the data (question 3). If students are unsure how to get started analyzing the data, suggest that they look in previous lessons for ideas. The amount of time students need to spend analyzing the data will depend on the scope of their questions. Encourage students to investigate their questions in depth and to pursue new questions. One question can lead to another—that's part of the data analysis process. If a question doesn't pan out, encourage students to come up with a new question to investigate.

Wrap-Up

5. Have students write conclusions or prepare brief presentations that answer these questions.

   - What question did you investigate?
   - What did you find out when you investigated your question?

- How did you investigate the question? What attributes did you look at? What kinds of plots did you make?
- How do you think your findings would be different if you investigated the same question for kindergartners? For adults?

6. As a final step, have students share their findings and graphs.
   - What new questions do your findings raise for you? What are your conjectures for your new questions?
   - What future studies could you plan to investigate your new conjectures?

ANSWERS

Investigating Data

1.–4. Answers will vary. You might want to use the Rubric for Making Comparisons as a guide for evaluating students' conclusions.

Box Plots

OVERVIEW

Some students have difficulty understanding box plots because the representation is abstract (no individual cases are shown) and looks so different from other graphs. In this lesson, students start with a familiar graph, a line plot, and transform it into a new graph, a box plot. This process helps students understand how data is represented in box plots. Students then use a five-number summary to construct a box plot.

Objectives

- Build understanding of how data is represented in box plots by transforming a line plot into a box plot
- Learn the terms *quartiles, interquartile range,* and *five-number summary*
- Compare line plots to box plots

TinkerPlots

Class Time: One class period

Materials

- Transform Line Plots into Box Plots worksheet (one per student)
- Make a Box Plot worksheet (one per student, *optional,* on CD)

Data Set: Music Time.tp (52 students estimated the amount of time they spent listening to music and doing homework in a typical week.), **Box Plots.tp** for demonstration

TinkerPlots Prerequisites: Students should be familiar with basic graphing; using dividers, percentages, median, and the **Drawing** tool.

TinkerPlots Skills: Creating box plots and hiding icons are explained in this lesson.

LESSON PLAN

Introduction

1. Introduce the lesson goal of learning about a new type of graph, box plots. Open the TinkerPlots file **Box Plots.tp,** which has a labeled box plot. Point out that box plots divide data into four equal parts and show where these divisions are. Explain that students are going to learn about box plots by using TinkerPlots to turn a line plot into a box plot.

2. Provide background information on the data set, **Music Time.tp.** 52 students estimated the amount of time they spend listening to music and doing homework in a typical week. Point out that this is a different data set from the one in Lesson 2.3.

Exploration

3. Have students work independently or in pairs to create a line plot of the attribute *Music_Time* and then turn it into a box plot (questions 2–6). Alternatively, if you have a computer projection system, you could go through the steps as a whole-class demonstration.

4. After students have transformed the line plot into a box plot, they compare the box plot they made with a line plot of the same music time data (questions 9–11). They use the five-number summary for the attribute *Homework_Time* to draw the box plot on their worksheet.

5. If you want students to have experience making box plots by hand and learn about five-number summaries, have them complete the Make a Box Plot worksheet from the *Digging into Data* CD.

Wrap-Up

6. If your students did the Make a Box Plot worksheet, have a whole-class discussion to compare the box plot for music time with the box plot for homework time.

   - What information did you need to know to draw a box plot for *Homework_Time*? [Five-number summary: minimum, lower quartile, median, upper quartile, maximum]
   - How does the amount of time this group of students tends to spend listening to music compare with the amount of time they spend doing homework? How can you tell from the box plots? [Students tend to spend a little more time listening to music than doing homework. All the parts of the box for music are to the right of the parts for homework, but the median is only a little to the right.]

     Note: Your students may point out that these students could be listening to music while they do their homework. This is a reasonable assumption; however, it is not possible to tell from the survey whether or not this is the case. You may want to ask students how they would change the survey question (see Lesson 2.1) to find out that information. For example, they could add a question like: How often do you listen to music when you're doing your homework? Choose 0 times per week, 1–2 times per week, 3–5 times per week, or 6–7 times per week. Another possibility is to change the original question to: How much time do you spend listening to music when you're not doing your homework?

7. Have a whole-class discussion about questions 9–11, which focus on comparing the two representations: line plots and box plots. Go over the questions and discuss which graph or graphs can be used to answer them. One important difference between the two representations is that line plots show individual cases and box plots do not. Box plots are useful for getting a big-picture view of the distribution of data when there are a lot of cases. (This lesson uses a small number of cases to help students manage the calculations and comparisons.)

ANSWERS

Transform Line Plots into Box Plots

9. a. 3 students; line plot. Information for individual students is available on the line plot but not on the box plot.
   b. 25%; both. It is easier to determine this percentage from the box plot because 18 is the upper quartile. Students may not realize that they can also determine this percentage from the line plot by counting the dots and doing calculations: $\frac{70}{70} = 25\%$.
   c. 0 h; both. Students can determine the fewest number of hours from both the box plot and the line plot. However, if the question asked how many students spend 0 h listening to music, then the answer would only be available on the line plot.

10. Sample answer: What is the highest number of hours that students listen to music? 24 h; both

11. Sample answer: How many students spend less than 10 h listening to music? 22 students; line plot

Make a Box Plot

1.

Attribute	**Minimum**	**Lower quartile**	**Median**	**Upper quartile**	**Maximum**
***Music_Time* (h)**	**0**	**8**	11	**18**	**24**
***Homework_Time* (h)**	4	7	10	14	22

2. Here is the box plot for *Homework_Time.*

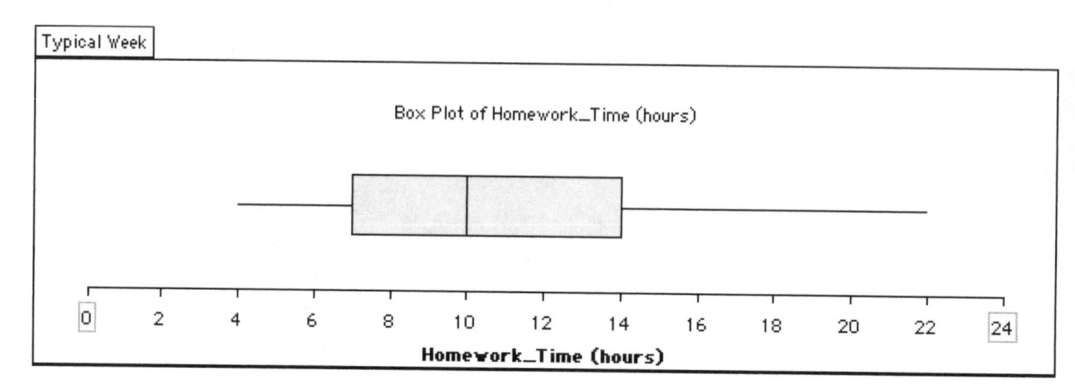

3. Sample statements:

- The median amount of time students spend listening to music is 11 h compared to a median of 10 h doing homework. This shows that this group of students tends to spend a little more time (1 h) listening to music than doing homework.
- The middle 50% of the students spend 8–18 h listening to music compared to 7–14 h doing homework. These interquartile ranges are similar, but the one for listening to music is a little larger, which shows that there is more variation in the amount of time that students typically spend listening to music.
- The left whisker for homework time is much shorter than the left whisker for music time. This shows that times in the lower quartile (or in the lower values) are more spread out for music time than for homework time.

Transform Line Plots into Box Plots

Name:

A box plot is a graph that summarizes how data are distributed. It divides the data into four equal parts and shows the division points.

Follow these steps to turn a line plot into a box plot.

1. Open the TinkerPlots file **Music Time.tp.** 52 students estimated the amount of time they spend listening to music.

2. Make a line plot of *Music_Time.*

3. To turn the line plot into a box plot, you'll divide the data into quarters. First use the median to divide the data in half.

4. Then use the dividers and percentages to divide the halves into quarters. To get four divisions, click the triangle next to the **Div.** button and choose **Number of Divisions** from the menu. Enter 4 and click **OK.**

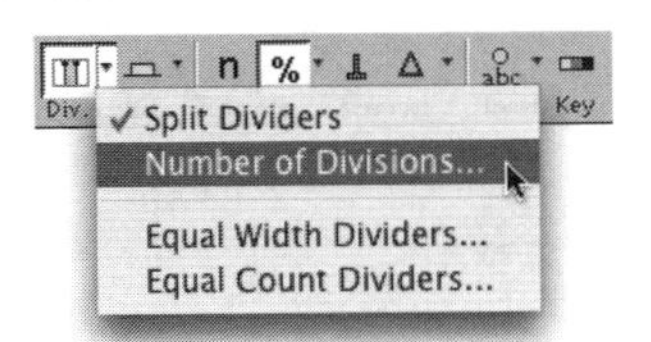

5. Use the **Drawing** tool to draw a box around the middle 50% of the data. The box goes from the *lower quartile* to the *upper quartile.* Show the location of the median by drawing a vertical line in the box.

6. To complete the box plot, add *whiskers* to both sides of the box. Draw the whiskers from the ends of the box to the lowest value (*minimum*) and highest value (*maximum*).

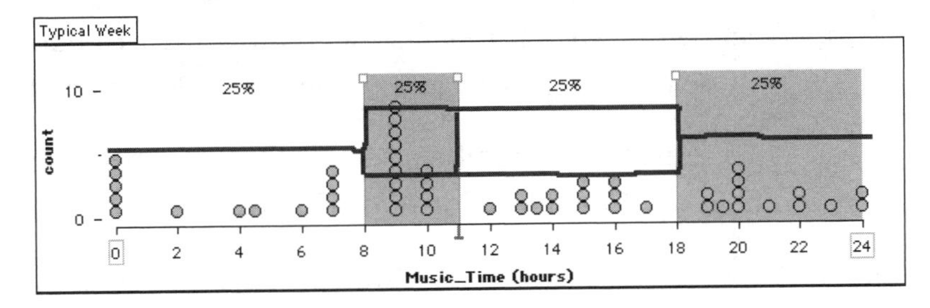

7. You can have TinkerPlots draw the box plot. Click the triangle next to the **Hat** button and choose **Box Plot** from the menu. The lines for the box plots should match the lines that you drew.

Hat
Percentile Hat
Range Hat
Average Deviation Hat
Standard Deviation Hat
Box Plot
Show Outliers

8. To see the box plot without the data icons, choose **Hide Icons** from the **Icon Type** menu on the lower plot toolbar

(where it says **Circle Icon**). Also turn off the dividers, percentages, and median.

9. Here is a line plot and a box plot of the same music time data. Answer each question and tell which graph or graphs could be used to find the answer.

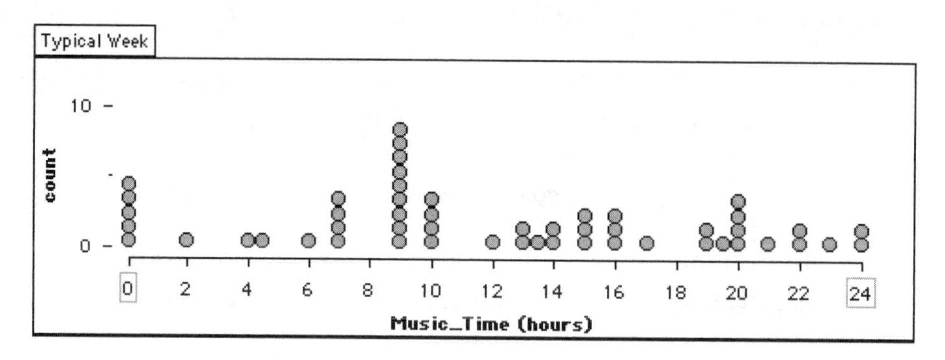

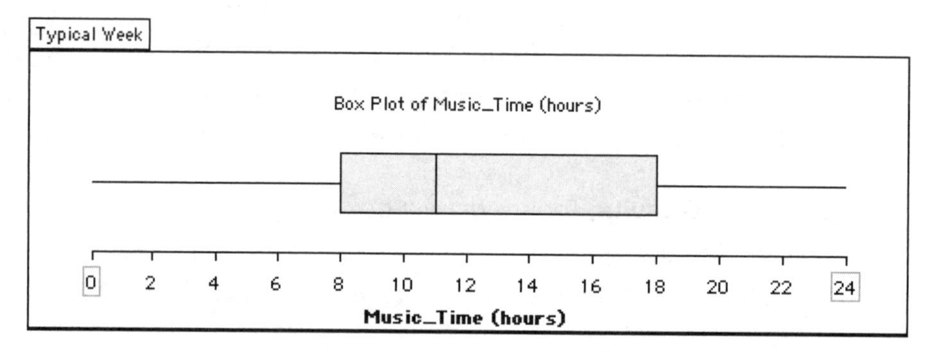

a. How many students spend 16 hours per week listening to music?
_________ Graph(s): Line plot Box plot Both

b. What percentage of students spend more than 18 hours listening to music? _________ Graph(s): Line plot Box plot Both

c. What is the fewest number of hours that students listen to music?
_________ Graph(s): Line plot Box plot Both

10. Write and answer one question that could be answered by using both graphs.

11. Write and answer one question that could be answered only by using the line plot.

Comparing Box Plots

OVERVIEW

Because box plots look so different from other graphs, students may have difficulty figuring out what to look at to make comparisons. This lesson provides a structured way for students to compare box plots, which involves examining specific features, such as the lengths of the whiskers, and then making interpretations in the context of the data.

Objectives

- Learn ways to compare box plots
- Interpret and compare box plots

Offline

Class Time: One class period

Materials

- Create Your Own Box Plot Reference Guide worksheet (one per student)
- Body Lengths and TV Time worksheet (one per student)
- Box Plots of Body Lengths transparency (*optional*, on CD)

LESSON PLAN

Introduction

1. To reinforce what students learned about box plots in Lesson 2.5, this lesson begins with a review of the key vocabulary terms. Students create their own reference guide by labeling an example box plot with the terms and by writing definitions for them (question 1). Students can use their work from Lesson 2.5 to help them create the reference guide. Explain that the example box plot shows data for the weights of 100 cats.

The round number of data points helps students find the quartiles.

2. Introduce ways to compare box plots by looking at the box plots for question 1 with the whole class. You might want to use the Box Plots of Body Lengths transparency from the *Digging into Data* CD. These box plots show the body lengths of 50 male and 50 female cats. Ask some concrete questions to help students become familiar with interpreting the data in this form.

- Lucky is a male cat that is 19 in. long. Which quartile contains the data for Lucky? [second quartile]
- About what percent of the female cats are 16 to 21 in. long? [50%]
- If you saw a line plot of the female cats' body lengths, in which quartile would you expect to see tall stacks of dots? Why? [We would expect to see tall stacks of dots in the second quartile because it is the narrowest. The values for about 13 cats are distributed across just 2 inches difference in body lengths (16–18 in.), so it makes sense that there are tall stacks of dots there.]

3. Discuss how to compare different features of the box plots. Ask students to first look for similarities or differences in the features, and then to interpret what this means in the context of cats' body lengths (question 1).
   - How do the lengths of the whiskers compare? What does that tell you?
   - How do the widths of the boxes (interquartile ranges) compare?
   - How do the positions of the boxes (interquartile ranges) compare?
   - How do the medians compare?
   - What are your conclusions about how the body lengths of male and female cats compare?

Exploration

4. Have students work in pairs to examine the other box plots to compare TV time for 61 male and 61 female students (question 2).

Wrap-Up

5. Use these questions for TV time as a way of checking for any difficulties students may have in reading box plots.
   - About what percent of male students spend more than 7 h watching TV? [75%]
   - About what percent of female students spend less than 15 h watching TV? [75%]
   - If you saw a line plot of the females' data, in which quartile would you expect to see dots that are the most spread out? [Fourth quartile] Dots that are the most densely packed together? [Second quartile]

6. Have a class discussion about what students found when they used the box plots to compare TV time for males and females.
    - What did you find when you compared TV time for males and females?
    - What do different parts of the box plots tell you about how the two groups compare?

ANSWERS

Create Your Own Box Plot Reference Guide

1. Sample answers:

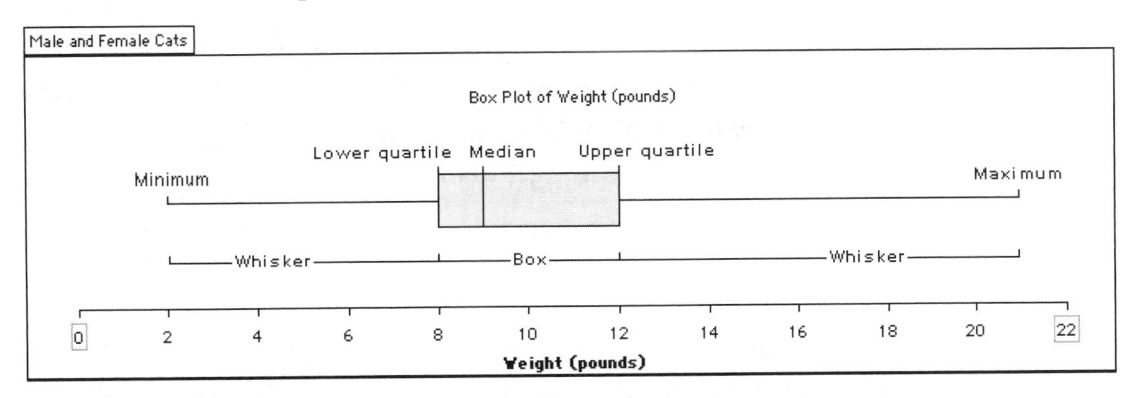

2. Students need to define four terms. Sample answers:

    The *box* contains about the middle 50% of the data.

    Each *whisker* contains either the lowest 25% or highest 25% of the data.

    The *median* divides the data set in half, so that about 50% of the values are above it and about 50% are below it.

    The *lower quartile* marks the lower end of the box; about 25% of the data are lower.

    The *upper quartile* marks the higher end of the box; about 25% of the data are higher.

    The *interquartile range* is from the lower quartile to the upper quartile.

    The *minimum* is the lowest value.

    The *maximum* is the highest value.

3. Answers will vary.

Body Lengths and TV Time

1. Sample answers:

Similarities and differences between the two box plots	What does this tell you about how the lengths of the male and female cats compare?
a. The whiskers on the left side of the box plots are shorter than the whiskers on the right side.	**The longest cats are farther from the median length than the shortest cats.**
b. **The interquartile range for females is wider than for males.**	**There is more variation in the length of the middle 50% of female cats than in the length of the middle 50% of male cats.**
c. **The box for the males is farther to the right and the median for males is 2 in. longer than for females.**	**Male cats tend to be longer.**

2.

Similarities and differences between the two box plots	What does this tell you about how the females' and males' TV times compare?
a. **Both box plots have longer whiskers on the right.**	**The students who watch the most TV are farther from the median time than the students who watch the least TV.**
b. **The box for males is wider.**	**There is more variation in the time the middle 50% of males spend watching TV than in the middle 50% of females.**
c. **The boxes start at about the same place, but the one for males extends farther to the right side.**	**The middle 50% of the males is more spread out than the females and extends to higher numbers of hours.**
d. **Median is higher for males.**	**Males tend to watch more TV.**

Create Your Own Box Plot Reference Guide

Name:

You will define the terms that describe parts of a box plot. Then you will use the terms as you interpret and compare box plots.

1. This box plot shows the weights of 100 cats. Label the box plot with each of these terms.

   Box *Whiskers* *Median* *Lower quartile* *Upper quartile* *Interquartile range*

   Minimum *Maximum*

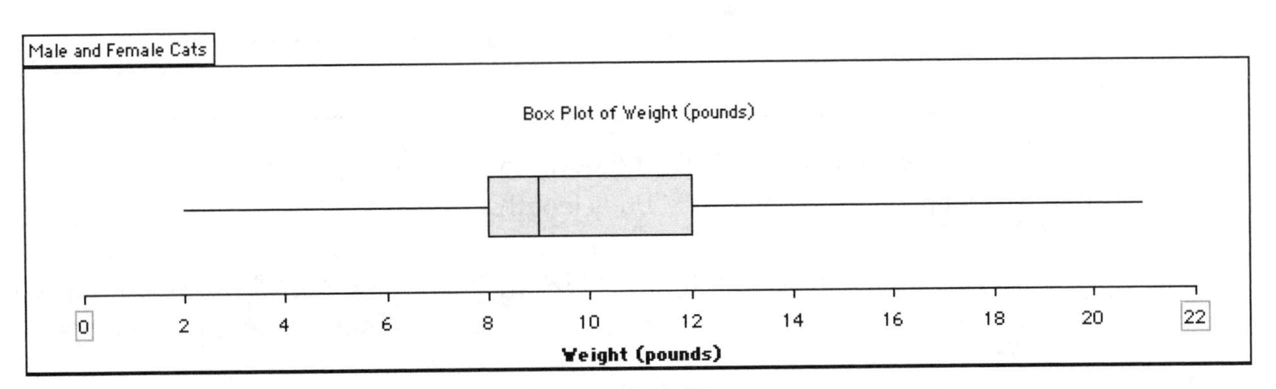

2. Pick four of these terms and write definitions for them.

3. Write some tips for making and interpreting box plots.

Body Lengths and TV Time

Name:

These box plots show the body lengths of 100 cats (50 male and 50 female).

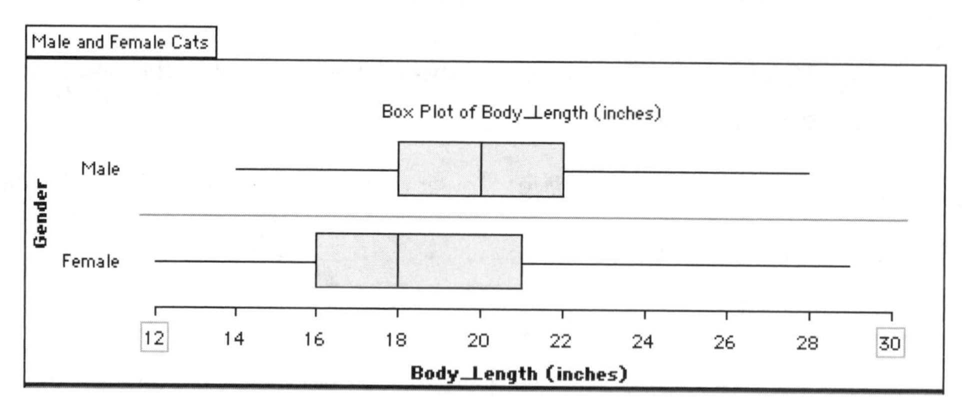

1. Compare the box plots to find out: How do male cats' body lengths compare with female cats' body lengths?

Similarities and differences between the two box plots	What does this tell you about how the lengths of the male and female cats compare?
a. How do the lengths of the whiskers compare? *The whiskers on the left side of both box plots are shorter than the whiskers on the right side.*	
b. How do the widths of the boxes (interquartile ranges) compare?	
c. How do the positions of the boxes (interquartile ranges) and the medians compare?	

These box plots show data for 122 students (61 male and 61 female) who estimated how much time they spend watching TV in a typical week.

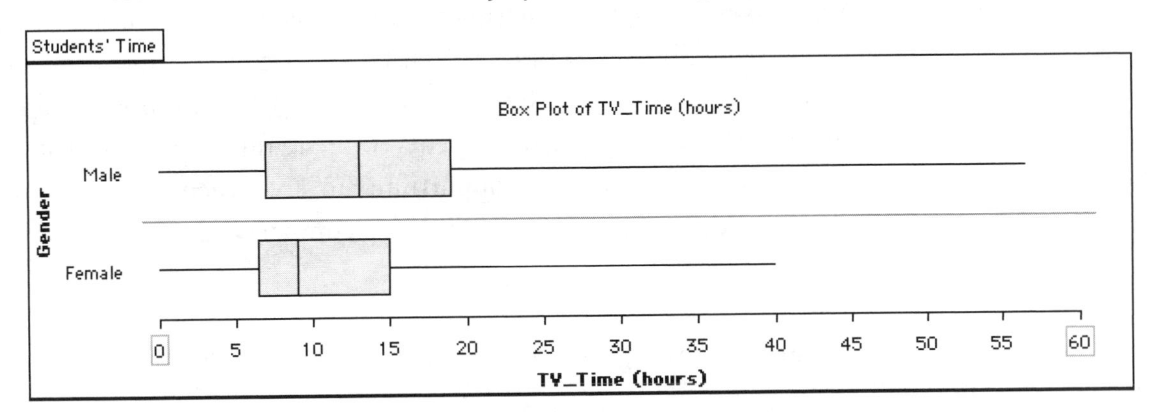

2. Compare the box plots to find out: How does the amount of time this group of females spend watching TV compare with the amount of time for this group of males?

Similarities and differences between the two box plots	What does this tell you about how the females' and males' TV times compare?
a. How do the lengths of the whiskers compare?	
b. How do the widths of the boxes (interquartile ranges) compare?	
c. How do the positions of the boxes compare?	
d. How do the medians compare?	

Comparing Three Groups

OVERVIEW

Students move from comparing groups of males and females to comparing sixth-, seventh-, and eighth-graders. This lesson offers more challenges than the previous lesson, by providing less structure for how to compare the box plots and by having students compare three box plots instead of two. This gives students the opportunity to strengthen and extend their understanding of box plots and making comparisons.

Objectives

- Compare three groups by using box plots
- Interpret and compare box plots

TinkerPlots

Class Time: One class period.

Materials

- Homework Time across Grades worksheet (one per student)
- Rubric for Making Comparisons (*optional,* on CD)

Data Set: Three Grades Time.tp (90 students—30 per grade—estimated the amount of time they spend doing homework in a typical week.)

TinkerPlots Prerequisites: Students should be familiar with basic graphing.

TinkerPlots Skills: Creating box plots and hiding icons are explained in this lesson.

LESSON PLAN

Introduction

1. Introduce the question: How does the amount of time spent doing homework in a typical week compare for sixth-, seventh-, and eighth-graders? Explain that the data set was collected from 90 students, 30 per grade, who estimated the amount of time they spend doing different activities in a typical week. In Lesson 2.6, students compared males and females. In this lesson, the focus shifts to comparing students in different grades.

2. Ask students to write their hypotheses individually before sharing them with the group. Students are likely to have different opinions about how homework time differs across the grades: Some students may say that eighth-graders spend the most time because they get the most homework, and other students may say that eighth-graders don't spend the most time because they are more efficient at doing their homework than sixth- and seventh-graders.

Exploration

3. Have students work individually or in pairs to compare the box plots for homework time.

   Troubleshooting tips for box plots:

   - If students choose **Box Plot** from the menu but nothing appears on the screen, they may not have fully separated the dots for *Homework_Time.*
   - If the grades are out of order, students can select a grade on the axis and move it to change the order.
   - If students are having trouble reading precise values, remind them to use reference lines.

4. If you have time, for an additional challenge have students do a similar comparison for another attribute of their own choosing (either Internet time or TV time).

Wrap-Up

5. Have a class discussion about the comparisons.

   - How did homework time compare across the grades? What evidence are you using to back up your conclusions?
   - If students did the extension: How did TV time compare across the grades? Internet time?
   - What parts of the box plots did you look at to make the comparisons?
   - How similar or different do you think the findings would be if we analyzed data from students in our school?

ANSWERS

Homework Time across Grades

1. Hypotheses and reasons will vary. Students will probably think students in higher grades spend more time doing homework.

6. Sample answers:

Similarities and differences across the box plots	What does this tell you about how the amount of time spent doing homework compares for sixth-, seventh-, and eighth-graders?
The whiskers on the left are shorter than the whiskers on the right for all three grades.	**The students who spend the most time doing homework are farther from the median time than the students who spend the least time.**
Sixth grade has the longest whisker on the right side.	**The sixth-graders who spend the most time doing homework are the farthest from the median.**
Seventh grade has the widest box (interquartile range), and it is positioned the farthest to the right (higher values on the scale).	**There is more variation in the time most seventh-graders spend, and they tend to spend more time than students in other grades.**
Seventh grade has the highest median: 18 h. That's 6 h higher than the medians for sixth and eighth grade.	**Seventh-graders tend to spend more time on homework. About half the seventh-graders spend more than 18 h on homework in a typical week. In comparison, about half the sixth- and eighth-graders spend more than 12 h per week.**
Sixth grade and eighth grade have the same median: 12 h.	**In both sixth and eighth grades, about half the students spend more than 12 h on homework and about half spend less.**

7. Sample answer: This group of seventh-graders tends to spend more time doing homework than these groups of sixth-graders and eighth-graders. The interquartile range shows that about 50% of the seventh-graders spend 12–24 h/wk doing homework. In sixth grade, students typically spend 7–15 h/wk on homework, and in eighth grade, students typically spend 8.5–16 h/wk. Seventh grade had the highest median, 18 h, which is 6 h higher than the median of 12 h for sixth grade and eighth grade.

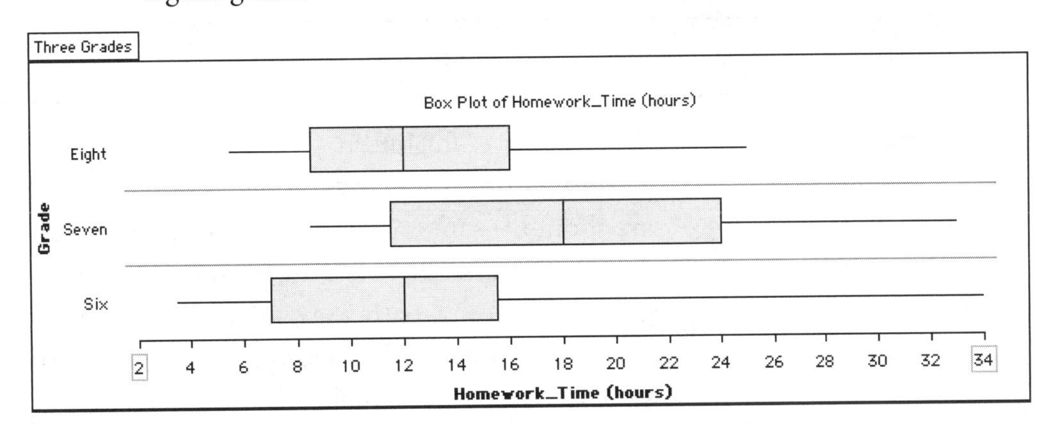

8. Sample answer for TV time: Eighth-graders tend to spend the least time watching TV during a typical week and to have the least variation in amount of time. Their median (8.3 h), upper quartile (13.0 h), maximum (25.0 h), and range (21 h) are all lower than the other grades. Sixth- and seventh-graders tend to spend similar amounts of time watching TV during a typical week. They have the same median (11.5 h). However, the times for the seventh-graders are more spread out than the times for the sixth-graders, and tend to be slightly higher. (Sixth grade has a range of 30 h; seventh grade has a range of 56.5 h. The center clump for sixth grade is 5.5–18 h, compared to 8–22 h for seventh grade.) Both the box and the whiskers are longer for seventh-graders. This means that overall the seventh-graders watched about the same amount of TV as the sixth-graders, but that the seventh-graders had more variation in their times, especially among those students who watched more than the median amount, 11.5 h.

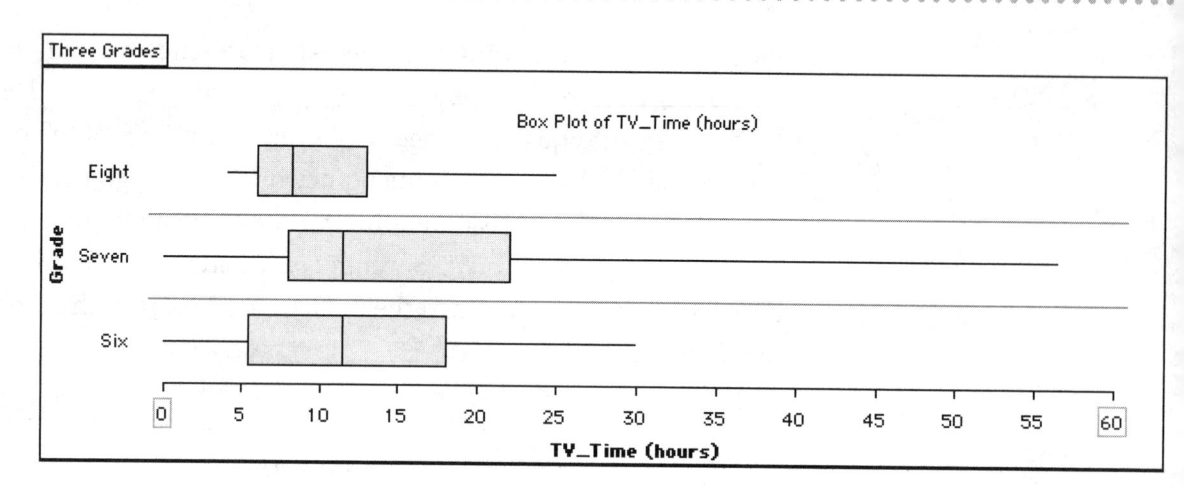

Sample answer for Internet time: The three grades are similar in the amount of time they tend to spend on the Internet. The median times are about the same (7 h for sixth and seventh grade, 8 h for eighth grade). There is more variation in Internet time for sixth and eighth grades than for seventh. (Seventh and eighth have similar ranges, 29.8 h and 27.5 h, and sixth has a range of 38 h. But the center clump for seventh (4.5–11 h) is smaller than the center clump for sixth (3.5–14 h) or eighth (5–15 h).) Eighth-graders generally spend more time on the Internet than sixth-graders, but some sixth-graders spend the most time of all, and the high-use sixth-graders are more spread out. The box for eighth grade is similar to the box for sixth grade but slightly to the right, but the right whisker for sixth grade is much longer (24 h compared to 12.5 h).

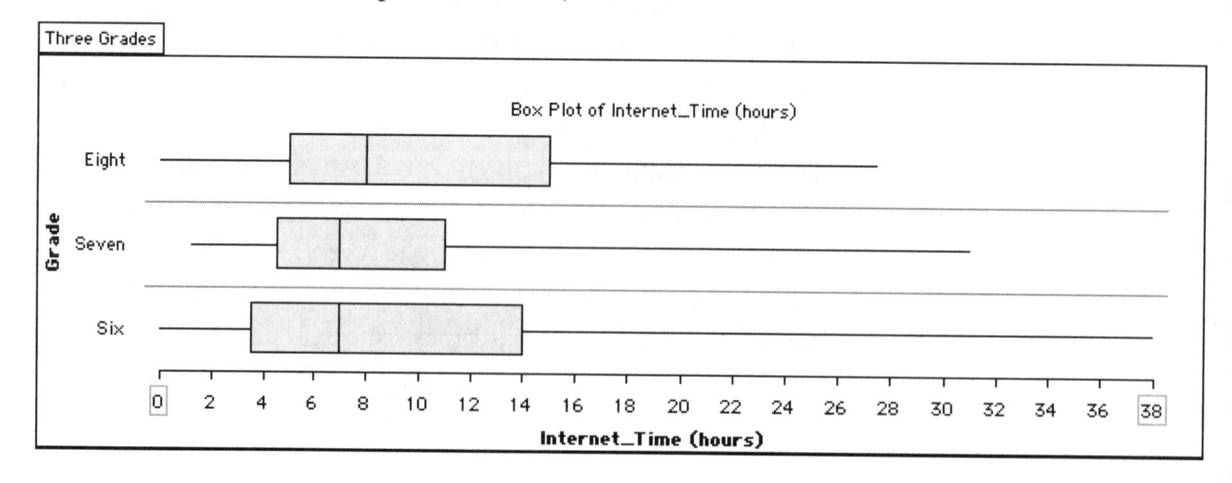

Homework Time across Grades

Name:

You will use box plots to compare homework time for three grades.

ASK A QUESTION AND MAKE A HYPOTHESIS

How does the amount of time spent doing homework during a typical week compare for sixth-, seventh-, and eighth-graders?

1. What do you think the data will show? Give reasons for your hypothesis.

ANALYZE DATA

2. Open the TinkerPlots file **Three Grades Time.tp.** 90 students (30 per grade) estimated the amount of time they spend doing homework in a typical week.

Follow these steps to make box plots to compare the amount of time spent on homework for the three grades.

3. Make a line plot of *Homework_Time.* Then put *Grade* on the other axis. Fully separate the data by *Grade* so that you have three line plots.

4. Choose **Box Plot** from the **Hat** menu. You should get one box plot for each grade.

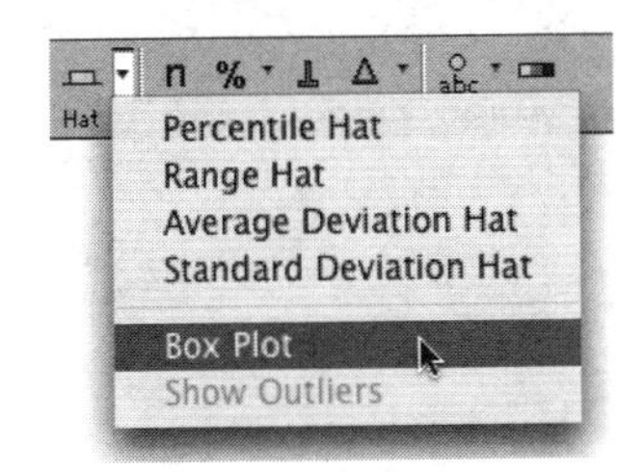

5. Choose **Hide Icons** from the **Icon Type** menu in the lower plot toolbar to remove the dots and see the box plots on their own.

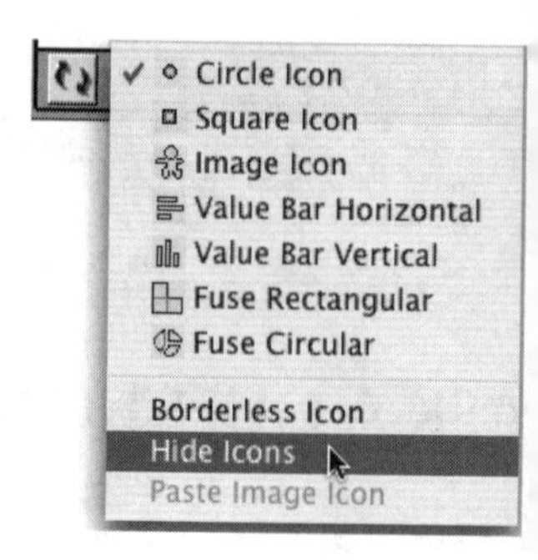

6. Compare the box plots.

Similarities and differences across the box plots	What does this tell you about how the amount of time spent doing homework compares for sixth-, seventh-, and eighth-graders?

COMMUNICATE CONCLUSIONS

7. What did you find out from the data? Did your conclusion support your hypothesis? Why or why not? Write your conclusion on a separate sheet of paper.

EXTENSION

8. Choose either TV time or Internet time and use box plots to compare the amount of time spent by students in the three grades. Keep your notes on a separate piece of paper.

Section 3

Comparisons Using Formulas: Investigating Data about Signatures and Words

This section extends the work students have done with comparing groups in Sections 1 and 2. Here, students learn more sophisticated methods to make comparisons, such as finding differences to compare measurements and using percentages to compare groups with different sizes. Students write formulas in TinkerPlots to perform calculations for all the cases in a data set. By using formulas, students create new attributes that enable them to analyze the data in more depth. This uses algebraic thinking. Students also begin exploring the relationship between two attributes by using a variety of graphical representations. Throughout the section, students use rubrics to guide them in writing strong, evidence-based conclusions.

OBJECTIVES

Collecting Data

- Collect data by taking measurements (Lesson 3.1)

Analyzing Data

- Develop an aggregate view of data for a group of cases (Lesson 3.1)
- Formulate questions and plan new studies (Lessons 3.1)
- Compare two groups with unequal sample sizes (Lesson 3.3)
- Create, interpret, and compare line plots and box plots (Lessons 3.1–3.4)
- Compare means, medians, ranges, and interquartile ranges (Lessons 3.2–3.4)
- Use formulas to create new attributes (Lessons 3.1–3.3)
- Explore whether there is a relationship between two attributes (Lesson 3.4)

- Create and compare a variety of plots for showing the relationship between attributes (Lesson 3.4)

Communicating about Data

- Describe and summarize data (Lesson 3.1)
- Write clear conclusions that use the data as evidence (Lessons 3.3, 3.4)
- Build understanding of the characteristics of high-quality conclusions (Lessons 3.3, 3.4)

Applications of Math Concepts from Other Strands

- Number and Operations: Apply knowledge of positive and negative integers (Lessons 3.1, 3.2)
- Number and Operations: Use percentages to analyze data (Lessons 3.1–3.3)
- Number and Operations: Write formulas to perform calculations for all the cases in a data set (Lessons 3.1–3.3)
- Measurement: Measure lengths in millimeters (Lesson 3.1)

TINKERPLOTS SKILLS

This section builds on the TinkerPlots skills learned in previous sections. Students should be comfortable with making line plots and box plots, using dividers, and displaying percentages, counts, medians, and means. In this section students learn how to change bin widths and axis endpoints and add text boxes to their files. Students also create new attributes and define them with formulas.

Introducing Data about Signatures

OVERVIEW

This lesson prepares students for working with the Signatures data set in Lessons 3.2 and 3.3. By measuring their own signatures, students build their understanding of how the data was collected and become familiar with the attributes. They learn how to create a new attribute by calculating the differences in lengths between their own signatures. This is the first time in the book that students work with an attribute that has negative values. Because this could cause confusion, it is important to clarify the meaning of negative values, zero values, and positive values in the context of the signatures. Students can use their own data as a concrete reference for building their understanding of these values and thus be better prepared for working with a larger, more abstract data set.

Objectives

- Learn about the attributes in the data set and how the data were collected
- Collect data by taking measurements
- Interpret data for a group of cases and communicate findings
- Compare data for different attributes (for the same group)
- Apply knowledge of positive and negative integers
- Formulate questions about data

Offline

Class Time: One class period

Materials

- Measure Your Signatures worksheet (one per student)
- Table of Differences in Signature Lengths transparency (*optional*, on CD)
- Millimeter rulers (one per pair of students)
- Sticky notes in three different colors (*optional*)
- Chart paper (*optional*)

LESSON PLAN

Introduction

1. Begin by having students sign their names in cursive on a piece of paper. Have them compare their signatures with their classmates in small groups. Alternatively, students could look at famous signatures, such as the signatures on the U.S. Declaration of Independence. These can be viewed and printed from the National Archives Experience website at www.archives.gov/national-archives-experience (choose Charters of Freedom, then Declaration of Independence). Discuss the different attributes of signatures.
   - What do you notice about the different signatures?
   - Which signatures stand out? Why?
   - How are these signatures similar to or different from your own signatures?
2. Discuss the meaning of *dominant* and *non-dominant hands.* Your dominant hand is the one that you usually use to write. If some students are ambidextrous, suggest that they call the hand that they usually write with the dominant hand.
3. Have students sign their names in cursive twice: first with their dominant hand and then with their non-dominant hand. Then, students measure the lengths of their signatures in millimeters (questions 1–3). Emphasize that students should measure each signature twice to make sure their measurements are as precise as possible.

   If students are unfamiliar with taking metric measurements, have them practice by first measuring the length of the blank lines on the worksheet in millimeters.

Millimeters are used for this activity to help students take precise measurements and compare small differences.

4. Have students compare the measurements for their signatures in small groups. This is an opportunity for students to look for values that stand out and may be caused by measurement errors. You may want to say something like: "Sometimes when people take measurements they can make errors in reading the ruler or recording the data. Let's check our data for potential errors. If you think there might be an error, you can re-take your measurements."

Exploration

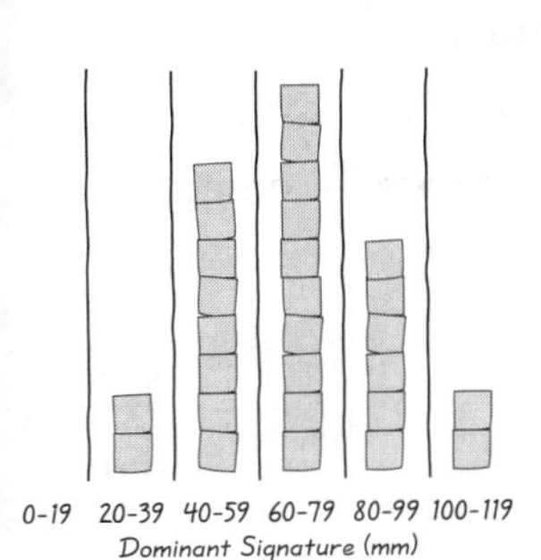

5. To collect the class data quickly and efficiently, have each student write the lengths of his or her signatures on sticky notes—use one color of sticky note for dominant signatures and a different color for non-dominant signatures. Use these sticky notes to put together plots of the class data on chart paper. Begin by making a class plot of the lengths of the students' dominant signatures. Because there will be many different measurements, it is helpful to group the data into intervals of 20 mm, such as 0–19, 20–39, 40–59, 60–79, 80–99, and 100–119 mm. Draw intervals on chart paper and have students add their lengths by putting their sticky notes in the appropriate interval or by drawing X's.

   Begin the class discussion of the plot by having students analyze the class's data on dominant signatures. Emphasize that students need to find out about the signatures for the class as a group and not for individual students.

   - What did you notice about the group of dominant signatures?
   - What are typical signature lengths for the class?
   - What do you think the plot of the non-dominant signatures will look like? How do you think it will be similar to or different from this one? Why?

6. Make a new plot for the data for non-dominant signatures using different colored sticky notes or markers. Use the same intervals so that students can compare the two plots (questions 5 and 6).

   - What did you notice about the group of non-dominant signatures?
   - What are typical non-dominant signature lengths for the class?
   - What similarities and differences do you notice between the data for the two signatures? What does that show about how the two kinds of signatures compare?

7. Students have been examining the lengths of the two signatures for the class overall. Shift the focus by introducing a new question: Do students tend to have non-dominant signatures that are longer, shorter, or the same length as their dominant signatures? Students may not realize that this question is different from the previous discussion question. Clarify that here we are interested in the differences in length for each student.

Introduce the idea of creating a new attribute called *Diff_Length* to compare the difference in signature lengths for each student. For example, this student's non-dominant signature is 80 mm and her dominant signature is 75 mm, so the difference in length (*Diff_Length*) is 5 mm.

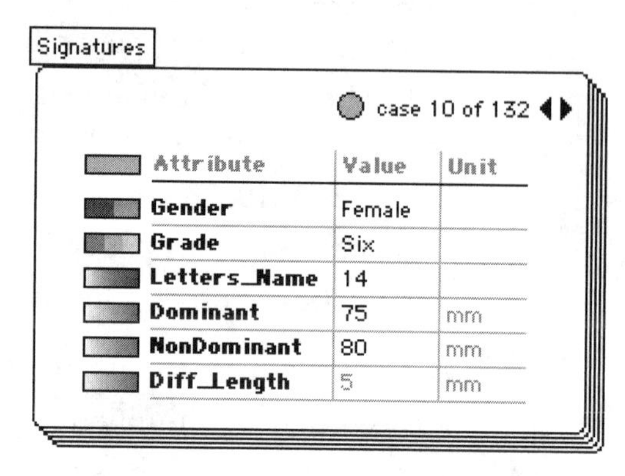

Note: Students could calculate *Diff_Length* by subtracting the length of the dominant signature from the length of the non-dominant signature or vice versa. It is important for all the students to use the same formula so that the class can discuss the results. Non-dominant signatures tend to be longer, so you might steer students to the formula *Diff_Length* = *NonDominant* – *Dominant.*

8. After students calculate their own *Diff_Lengths*, collect the data and make a separate class plot using a third color of sticky notes or marker. Use the discussion questions to build students' understanding of the attribute.

   - What is the shortest *Diff_Length* in our class? The longest *Diff_Length*?
   - What does it mean if a student's *Diff_Length* is 0? If a student's *Diff_Length* is a positive integer? If a student's *Diff_Length* is a negative integer?

Wrap-Up

This table is available as an overhead transparency.

9. Make a table to organize the class's data for the attribute *Diff_Length* (question 6).

10. Have a class discussion about the table.

    - Do students in our class tend to have non-dominant signatures that are longer, the same size, or shorter than their dominant signatures?

- What evidence are you using from the table to support your conclusions? What evidence could you use from the graph to support your conclusions?
- What might be the reasons for these findings?
- What future studies would you like to conduct about dominant and non-dominant hands? How would you plan those studies?

ANSWERS

Measure Your Signatures

Answers will vary because they depend on the data that the class collects. Students' non-dominant signatures tend to be longer than their dominant signatures. This may be because students have less control with their non-dominant hands, so their writing is sloppier and larger.

5. b. A *Diff_Length* that is a positive integer means their non-dominant signature is longer than their dominant signature.

   c. A *Diff_Length* that is a negative integer means their dominant signature is shorter than their non-dominant signature.

Measure Your Signatures

Name:

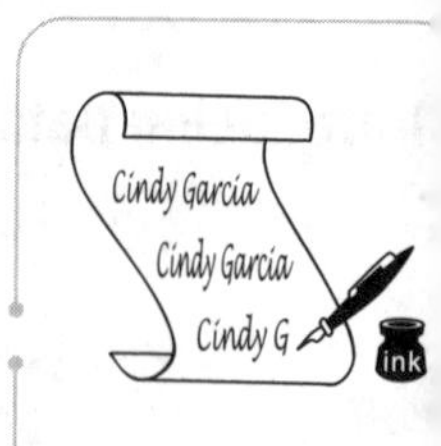

Follow these steps to collect data on your signatures.

1. Your *dominant* hand is the one you usually use. Which hand do you use to write?

   Right Left

   Use that hand to sign your first and last names in cursive on the line.

   ____

2. Your other hand is your *non-dominant* hand. Use that hand to sign your first and last names in cursive.

   ____

3. What is the length of each signature in millimeters? Measure twice to make sure that your measurements are precise. Add your information to the empty data card below. You can read about the attributes below the data card.

4. What is the difference in length between your non-dominant signature and your dominant signature? This attribute is called *Diff_Length: Diff_Length = NonDominant – Dominant.*

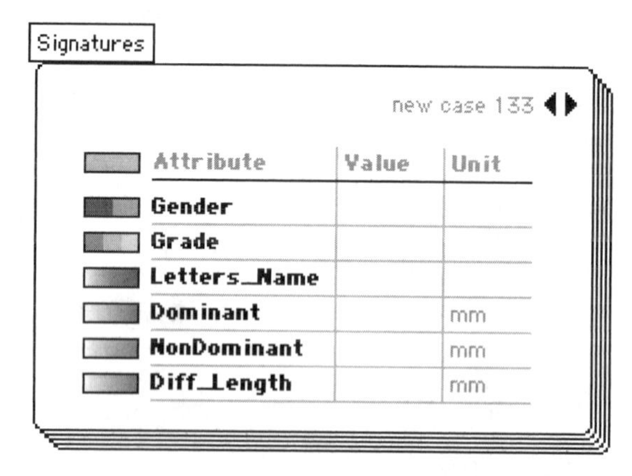

Signatures

new case 133

Attribute	Value	Unit
Gender		
Grade		
Letters_Name		
Dominant		mm
NonDominant		mm
Diff_Length		mm

Letters_Name: Number of letters in first and last names (does not include middle name)

Dominant: Length of signature written with dominant hand

NonDominant: Length of signature written with non-dominant hand

Diff_Length: Difference in lengths between the signatures

5. *Diff_Length* can be positive, 0, or negative.

   a. Is your *Diff_Length* a positive integer, 0, or a negative integer?

   ____

   b. What does it mean if a student's *Diff_Length* is a positive integer?

   c. What does it mean if a student's *Diff_Length* is a negative integer?

6. Organize the class's data for the attribute *Diff_Length* in the table.

	Diff_Length		
	Negative integer	**0**	**Positive integer**
Number of students			
Percent of students			

7. Do students in your class tend to have non-dominant signatures that are longer, the same size, or shorter than their dominant signatures? Use evidence from the table to support your conclusions.

Using Differences to Compare Two Attributes

OVERVIEW

In this lesson, students use a formula to create a new attribute to compare the differences in lengths between two kinds of signatures. The values for this attribute include positive and negative integers, so students need to apply their knowledge of integers in a data analysis context. To help build students' understanding of the attribute, they first find individual cases that have specific positive or negative values and then progress to analyzing the group of cases. Students also have the opportunity to apply a variety of data analysis methods used in previous lessons, including representing data in intervals and line plots.

Objectives

- Use a variety of methods to make comparisons
- Create and compare graphs, including line plots
- Use a formula to create a new attribute
- Apply knowledge of positive and negative integers

TinkerPlots

Class Time: One class period

Materials

- Dominant and Non-dominant Signatures worksheet (one per student)

Data Set: Signatures.tp (132 middle-school students)

TinkerPlots Prerequisites: Students should be familiar with intermediate graphing and adding a formula-derived attribute.

TinkerPlots Skills: Duplicating a plot and changing bin widths are explained in this lesson.

LESSON PLAN

Introduction

1. Introduce the question: How do the lengths of signatures written with non-dominant hands compare with the lengths of signatures written with dominant hands?

   Point out how this investigation builds on the previous lesson. In that lesson, students looked at a small sample of data that they collected.

In this lesson, they will analyze a large data set (132 cases) that was collected by other students, and they will be able to use TinkerPlots to create different kinds of plots.

2. Encourage students to consider how the findings for this larger data set will be different from the findings for their own class. Have students write their hypotheses individually and then share them with the group (question 1).

Exploration

3. Students work independently or in pairs to investigate the data. They will use a formula to create a new attribute, *Diff_Length*, and will create different plots (questions 3–7).

   Note: As you circulate around the room, check that all the students are using the formula *Diff_Length* = *NonDominant* – *Dominant*. It's important that students all use the same formula to avoid confusion later in the class discussion.

4. If some students have completed the investigation and need an additional challenge, have them explore the relationship between the two kinds of signatures. Pose these questions.

   - Do students with long dominant signatures tend to have long non-dominant signatures?
   - Do students with short dominant signatures tend to have short non-dominant signatures?

   Students can make line plots for *Dominant* and *NonDominant* and use highlighting to find individual students in both graphs. (To turn off the color, click the blue box beside the Attribute column head in the data cards. Then click an icon to highlight it in all plots.) Students could also color both plots by *Dominant,* for example, and compare the coloring. Does the *NonDominant* plot show about the same color gradient, implying that the students with long dominant signatures also have long non-dominant signatures, or are the colors mixed up, implying that they don't?

Wrap-Up

5. Have students write their conclusions about how the lengths of the two signatures compare (question 8). If time is short, have students make a list of their ideas so that they can write them up for homework.

6. Have a class discussion about the conclusions.
   - What did you find out about how the two kinds of signatures compare?
   - What did you find out by creating a new attribute, *Diff_Length*?
   - How do your findings for this data set compare with what we found out from our class data set?
7. If students did the extension on the relationship between the two attributes, then have a class discussion about their findings.
   - Do you think there is a relationship between the two attributes?
   - If you knew the length of a student's dominant signature, could you accurately predict the length of their non-dominant signature? Why or why not?

ANSWERS

Dominant and Non-dominant Signatures

1. Students may guess that non-dominant signatures will be longer, as their writing is less controlled with the non-dominant hand.

4. Students will probably use dividers, the median, and percentages to find the middle 50% of values, as they learned in Sections 1 and 2. Sample answer: The non-dominant signatures tend to be longer than the dominant signatures, and their length is more varied. The middle 50% of values for the dominant signatures are about 45–72 mm, with median 57 mm. The middle 50% of non-dominant values are about 50–92 mm, with median 72 mm.

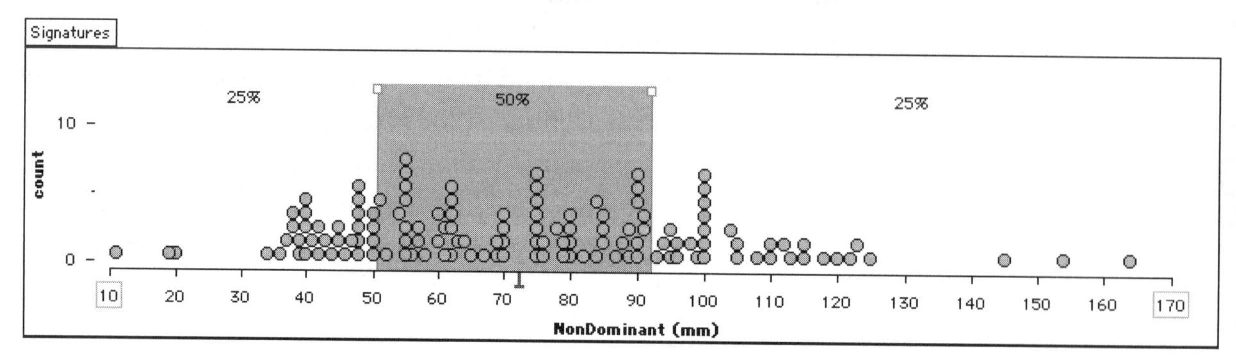

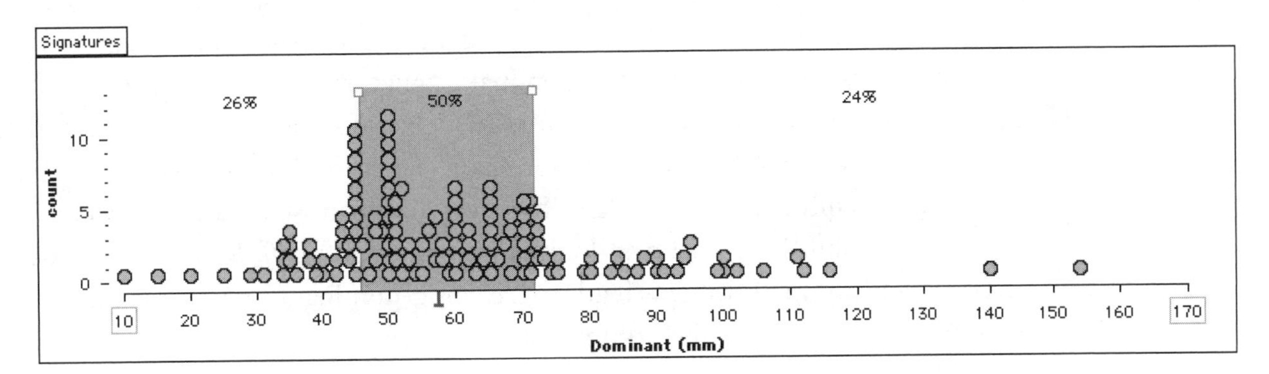

6. There are seven students with a *Diff_Length* of 20 and three students with a *Diff_Length* of 0. Students only need to find one of each.

Diff_Length (mm)	Case #	Non-dominant signature length (mm)	Dominant signature length (mm)
20	13	100	80
	18	51	31
	36	104	84
	73	55	35
	102	62	42
	107	60	40
	108	90	70
0	7	40	40
	65	34	34
	95	90	90
–20	122	40	60
–10	71	46	56

7. a.

	Diff_Length		
	Negative integer	0	Positive integer
Percent of students	23%	2%	75%

b. Sample answer: 1–20 mm

8. Sample answer: 75% of the students had non-dominant signatures that were longer than their dominant signatures (positive *Diff_Lengths*). In comparison, fewer students (23%) had non-dominant signatures that were shorter than their dominant signatures (negative *Diff_Lengths*). Only 2% of the students had non-dominant signatures that were the same length as their dominant signatures (a *Diff_Length* of 0). This group of students tends to have non-dominant signatures that are longer than their dominant signatures.

9. a. Students' conclusions should incorporate their answers to questions 5, 9, and 10.

   b. Answers will vary.

10. In general, students with long dominant signatures tend to have medium to long non-dominant signatures, students with short dominant signatures tend to have short to medium non-dominant signatures, and students with medium dominant signatures tend to have medium non-dominant signatures. One way to approach this question is to use dividers and percentages to find the middle 50% in each graph, which divides the graph into short, medium, and long signatures. Then, highlight each group in one graph and find where those cases are in the other graph. To view only the highlighted cases, choose **Hide Unselected Cases** from the **Plot** menu. (To keep the cases highlighted in the other plot as you select it, make sure you click on the plot toolbar, not in the actual graph.) Then reposition the dividers to analyze only the cases showing.

    These plots show the non-dominant lengths of the long dominant signatures (those in the top 25%, above 70 mm) compared to all the non-dominant lengths. 66% of the cases with long dominant signatures also have long non-dominant signatures (92.2 mm or above).

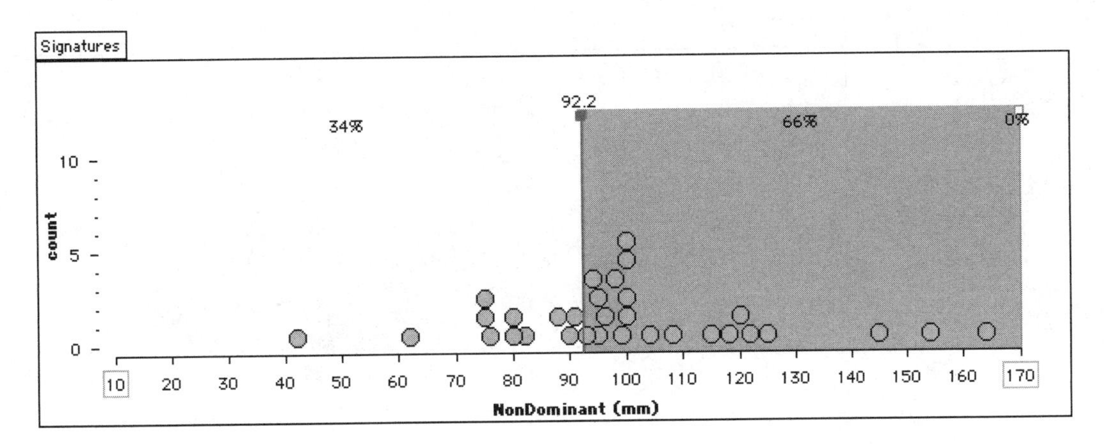

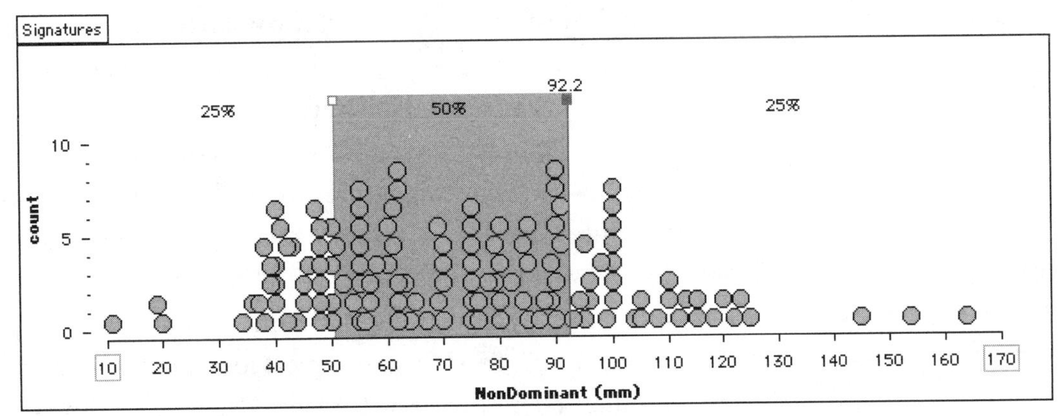

These plots show the non-dominant lengths of the short dominant signatures (those in the bottom 25%, below 45.9 mm) compared to all the non-dominant lengths. 49% of the cases with short dominant signatures also have short non-dominant signatures (50.5 mm or below), but 51% have medium non-dominant signatures.

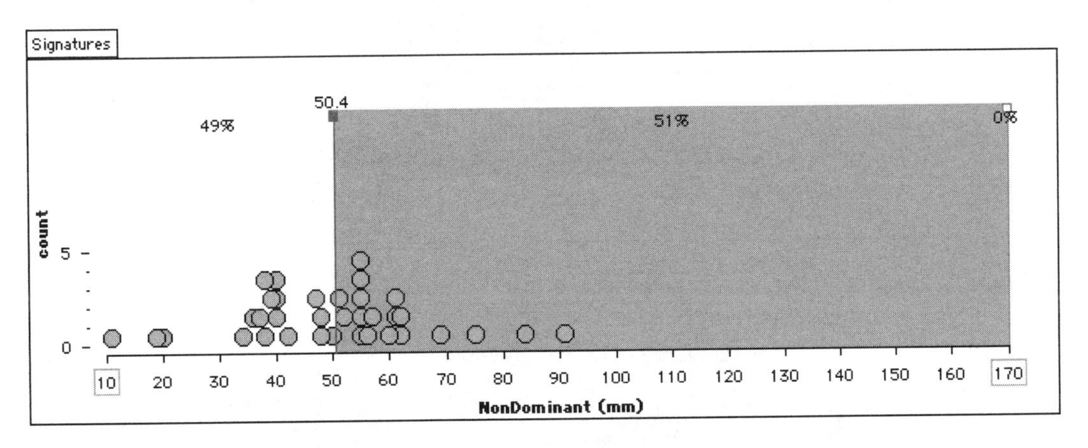

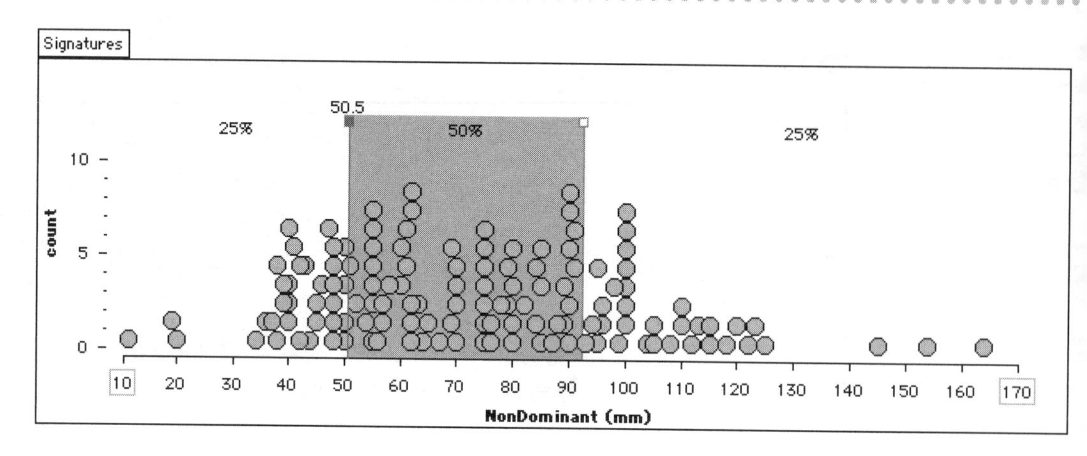

11. Sample answer: I don't think if you knew the length of a student's dominant signature, you could predict the length of their non-dominant signature. The graphs show non-dominant lengths for each group of dominant lengths (short, medium, and long). The ranges of the non-dominant signatures for each group of dominant signatures are quite large (short: 90 mm; medium: 84 mm; and long: 122 mm). Typical short dominant signatures have non-dominant lengths of 40–57 mm, typical medium dominant signatures have non-dominant lengths of 55–89 mm, and typical long dominant signatures have non-dominant lengths of 88–108 mm. Each group shows a lot of variability, and the typical lengths overlap, so I don't think you could tell.

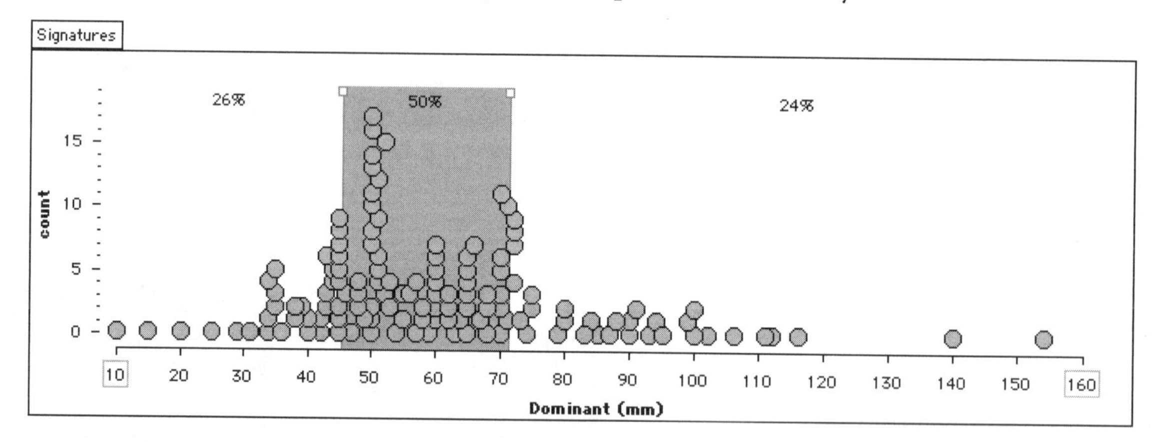

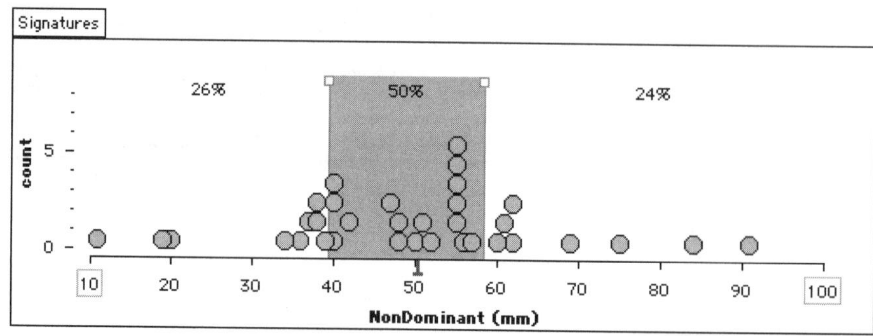

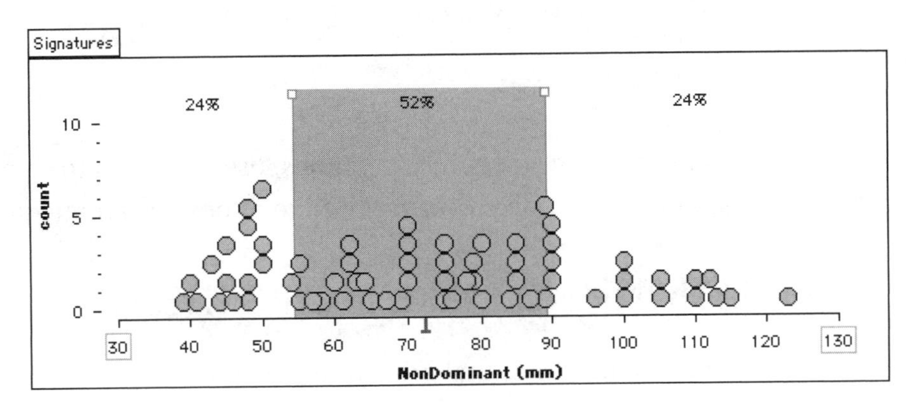
Signatures
24%
52%
24%
10
5
0
count
30
40
50
60
70
80
90
100
110
120
130
NonDominant (mm)

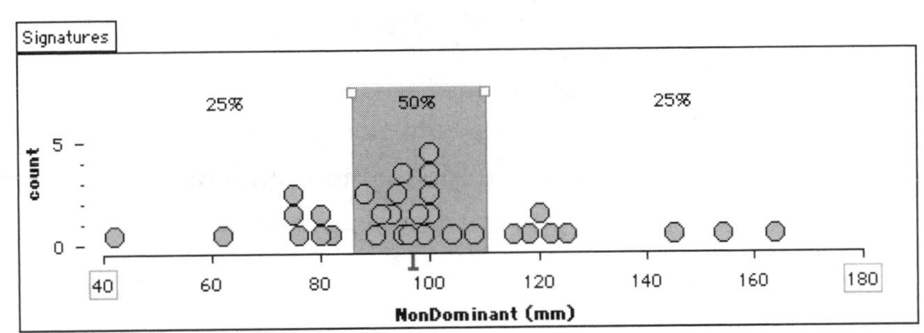
Signatures
25%
50%
25%
5
0
count
40
60
80
100
120
140
160
180
NonDominant (mm)

Dominant and Non-dominant Signatures

Name:

You will create and compare graphs, use a formula to create a new attribute, and apply your knowledge of positive and negative integers.

ASK A QUESTION AND MAKE A HYPOTHESIS

How do signatures that are written with non-dominant hands compare with signatures written with dominant hands? Are they longer, shorter, or about the same?

1. You analyzed this question for your own class. Now, you will look at data from 132 middle-school students. What do you think the data for this group of students will show? Give reasons for your hypothesis.

ANALYZE DATA

2. Open the TinkerPlots file **Signatures.tp.** 132 students signed their names two times: first with their dominant hands and then with their non-dominant hands. They measured their signatures in millimeters.

Compare Line Plots

3. Create two line plots: one for non-dominant signatures and one for dominant signatures. To make it easier to fairly compare the data, change the scales for both graphs to 10–170 mm. You can change the starting and ending values by double-clicking the boxes at the ends of the scale.

4. How do the lengths of the two kinds of signatures compare?

Use a Formula to Compare Lengths

5. Use TinkerPlots to make a formula to calculate the differences in length for the two kinds of signatures. This new attribute is called *Diff_Length.*

Diff_Length = *NonDominant* – *Dominant*

6. Make a line plot for the attribute *Diff_Length.* Find one student with each value for *Diff_Length.*

Diff_Length **(mm)**	**Case #**	**Non-dominant signature length (mm)**	**Dominant signature length (mm)**
10	94	114	104
20			
0			
–20			
–10			

7. Analyze the data for the group of students.

   a. Use percentages and dividers to help you fill in the table.

	Diff_Length		
	Negative integer	**0**	**Positive integer**
Percent of students			

   b. What are typical *Diff_Length* values for this group of students?

   ____

8. What did you find out from your analysis of the attribute *Diff_Length*? What does this tell you about how the two kinds of signatures compare?

COMMUNICATE CONCLUSIONS

9. On a separate sheet of paper, write your conclusion for each question. Make sure to use the data as evidence to support your conclusions.

   a. How do the lengths of signatures written with non-dominant hands compare with the lengths of signatures written with dominant hands? How can you tell from the data?

   b. How does the data from this group of 132 students compare with the data from your class?

EXTENSION

Write your conclusions for these questions on a separate sheet of paper.

10. What is the relationship between the two kinds of signatures?

    a. Do students with long dominant signatures tend to have long non-dominant signatures? How can you tell from the data?

    b. Do students with short dominant signatures tend to have short non-dominant signatures? How can you tell from the data?

    c. How did you use TinkerPlots to investigate these questions?

11. If you knew the length of a student's dominant signature, could you accurately predict the length of his or her non-dominant signature? Why or why not?

Comparing Two Groups of Different Sizes

OVERVIEW

This lesson extends the work that students have done comparing groups. The new challenge is that the two groups have unequal sample sizes. Some students may feel that it is unfair or not possible to compare these groups. The lesson is designed to help students understand that when group sizes are unequal, comparing counts is misleading but comparing percentages is fair. Students also use a formula to create a new attribute to determine the average (mean) length per letter in the signatures. This new attribute enables students to compare signatures for names that have different numbers of letters.

Objectives

- Compare two groups of unequal sizes
- Understand that percentages are useful for making comparisons when group sizes are unequal
- Use a formula to create a new attribute
- Write conclusions by using the data as evidence

TinkerPlots

Class Time: One class period

Materials

- Signatures of Males and Females worksheet (one per student)
- Rubric for Making Comparisons (one per student, on CD)

Data Set: Signatures.tp (132 middle-school students: 59 boys and 73 girls)

TinkerPlots Prerequisites: Students should be familiar with intermediate graphing, adding a formula-derived attribute, and making box plots.

LESSON PLAN

Introduction

1. Introduce the question: How do the lengths of dominant signatures for this group of middle-school males compare with the lengths of dominant signatures for this group of middle-school females?
2. Have students come up their own hypothesis individually and then share with the group (question 1).

3. The first data analysis task (question 3) involves figuring out how to fairly compare the group of males (59) with the group of females (73). This is the first time that students need to compare two groups of unequal sizes.

   The task begins with example graphs and conclusions. If you have a computer projection system, go through this task as a whole class activity by demonstrating the graphs in TinkerPlots and leading a class discussion. For Darleen's graph, explain that the percentages are determined for each row. Point to 34% and ask, "Is this 34% of all the students, 34% of the 0–49 bin, or 34% of the males?" [34% of the males.] If you add up the percentages across the row, they will add up to about 100%. The total is not exactly 100% because of rounding (males: 101%; females: 99%). Students can easily get confused when displaying multiple systems of percentages. Point to different percentages and ask the class to say what each percentage represents. For example, 34% of the males have dominant signatures 0–49 mm long. Note that TinkerPlots can also display column percentages or cell percentages. You may want to show those (click the arrow beside the % button) and discuss what they mean, but this might add confusion.

4. After students are familiar with how TinkerPlots displays percentages, shift the focus to the question of why Tom and Darleen came to different conclusions (question 5).

   - Let's start by looking at the first interval, 0–49 mm. Tom's graph shows that 20 males and 20 females had signatures that were 0–49 mm long. Darleen's graph also shows 20 males and 20 females, but the percentage of males (34%) is greater than the percentage of females (27%). Why is that? [There are more females, so 20 dots is a smaller percentage of the total number of females than it is of males.]
   - Tom and Darleen divided the data into the same intervals. Why did they come up with different conclusions? [Tom compared counts and Darleen compared percentages.]
   - Which conclusion do you think is a fairer comparison of the two groups? Why? [Comparing counts is misleading because there are more dots for the females than for the males. To fairly compare the two groups you need to use percentages.]

Note: If students are confused about the need to use percentages, it may be helpful to give an example from another context. For example, 80 seventh-grade students and 60 eighth-grade students each got a turn to shoot a basket. 40 seventh-graders scored a basket and 36 eighth-graders. Which grade performed better? How can you fairly compare the two grades? By using percentages you can compare the percentage of students in each grade that scored a basket—50% of the seventh-graders compared to 60% of the eighth-graders. Comparing the counts (40 to 36) would be misleading because more seventh-graders had a chance to shoot a basket.

Exploration

5. After discussing the importance of using percentages, students need to consider how the number of letters in names may affect the lengths of signatures. If females tend to have more letters in their names than males then it is likely that their signatures would be longer. Students can create a new attribute, *Letter_Length,* to have TinkerPlots calculate the average length per letter in each student's signature.

$$\mathit{Letter\_Length} = \frac{\mathit{Dominant}}{\mathit{Letters\_Name}}$$

Some students may be troubled that the attribute doesn't take into account the space between first and last names. They could consider the space equal to one letter and add that to their formula.

$$\mathit{Letter\_Length} = \frac{\mathit{Dominant}}{\mathit{Letters\_Name} + 1}$$

This lesson assumes that students have had experience using TinkerPlots to create formulas. If they have not, demonstrate how to create the formula.

a. In the data card, double-click <**new attribute**> and type `Letter_Length`.

b. Expand the data cards until you can see the Formula column and double-click the formula circle for *Letter_Length.* A formula editor appears.

If you are using the second formula, you need to use parentheses in the denominator.

c. Enter the formula for *Letter_Length.* You can type the attribute names or expand the Attributes list and double-click the names there.

d. Click **OK.** The average letter length should appear in each data card.

e. Enter the units for the attribute in the Units column: `mm/letter`.

6. Students create different plots to compare the new attribute, *Letter_Length,* for males and females. The worksheet suggests that they use intervals, line plots, and box plots (question 6). If students do not have experience with box plots from Section 2, you can cut the box plots suggestion in this lesson.

Wrap-Up

7. Have students write their conclusions by following the criteria on the Rubric for Making Comparisons. If students did Lesson 1.7 or 2.4, point out that this is the same rubric that they used. If students are new to using the rubric, then you may want to wait to introduce it during an offline class period so that you can spend time going over the characteristics of high-quality conclusions. Have students make a list of their findings that they can use later to write their conclusions.

8. Have a class discussion about the conclusions.

   - What did you find out by analyzing the new attribute, *Letter_Length*?
   - In Lesson 3.2 you created the attribute *Diff_Length,* which had positive and negative values. Is it possible to get negative values for *Letter_Length*? Why or why not?
   - Did your findings support your hypothesis? Why or why not?
   - What other questions about signatures would you like to investigate? How would you collect and analyze the data?

9. As an extension, have students analyze the same data set to investigate a new question: How do the lengths of non-dominant signatures compare for males and females?

ANSWERS

Signatures of Males and Females

Note: These answers use the formula $Letter\_Length = \frac{Dominant}{Letters\_Name}$

1. Hypotheses and reasons will vary.

3. b. Tom was comparing counts and Darleen was comparing percentages. There are the same numbers of males and females in the lowest interval, but different percentages. In both cases there were more females in the middle two intervals, but using counts, Tom found 15 more, where Darleen found only 9% more.

   c. Students should realize that Darleen's method of using percentages is a good way to fairly compare the two groups. When the groups are different sizes, comparing percentages makes more sense than comparing counts.

4. a. 6 mm. Sample answer: I divided the length of Marina's dominant signature, 60 mm, by the number of letters in her name, 10.

   b. Answers will vary.

6. Sample answers:

   a. The intervals plots show that most students, 89% of males and 91% of females, have letter lengths between 3 and 8.999 mm/letter. This means males' and females' letters have similar lengths in general.

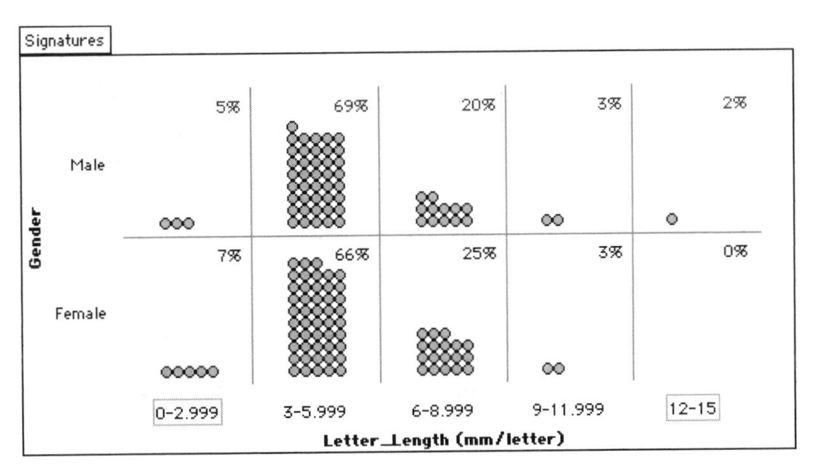

b. The line plots show that females typically have very slightly higher values for letter length than males. The middle 50% of females' values is from 4.12 to 6.13 mm/letter, while the middle 50% of males' values is from 3.75 to 6.00 mm/letter. The females' median (5 mm/letter) is also slightly higher than the males' (4.8 mm/letter). So females typically write their letters slightly larger or longer than males.

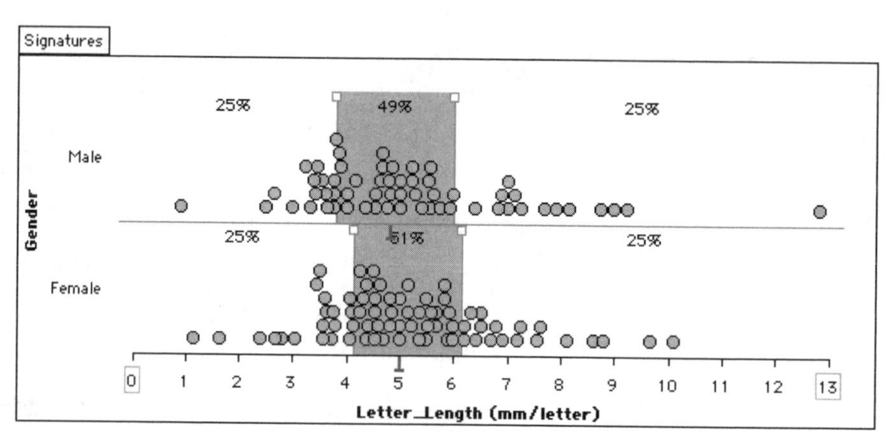

c. The box plots show that females typically have very slightly higher values for letter length. (See 10b for details.) One feature that is most apparent in the box plots is that males have a longer range of letter lengths: 11.92 mm/letter, compared to a range of 8.94 mm/letter for females. So females typically write their letters slightly larger or longer than males, but males vary more in the size of their letters.

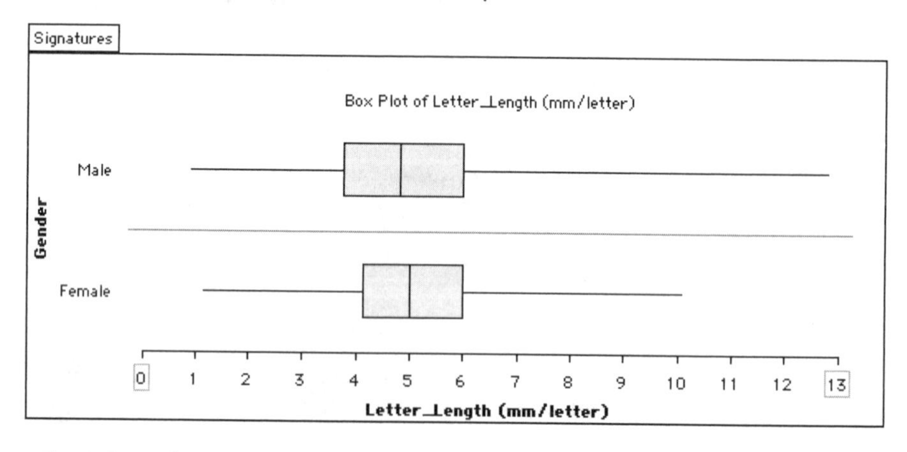

d. Other plots used will vary.

7. Students' conclusions should include their observations from question 6 and follow the Rubric for Making Comparisons.

Signatures of Males and Females

Name:

You will compare two groups with different sizes, and use a formula to create a new attribute.

ASK A QUESTION AND MAKE A HYPOTHESIS

How do the lengths of dominant signatures compare for middle-school males and middle-school females?

1. What do you think the data will show? What are your reasons for this hypothesis?

ANALYZE DATA

2. Open the TinkerPlots file **Signatures.tp.** 132 students in grades 6–8 measured their signatures in millimeters.
3. Two students, Tom and Darleen, analyzed the data by dividing it into equal intervals.
   a. Look at Tom's and Darleen's graphs on the next page and make them yourself with TinkerPlots.
   b. Read Tom's and Darleen's conclusions. Why did they come up with different conclusions?
   c. Which conclusion do you think is a fairer comparison of the two groups? Why?

Tom's graph and conclusions	Darleen's graph and conclusions
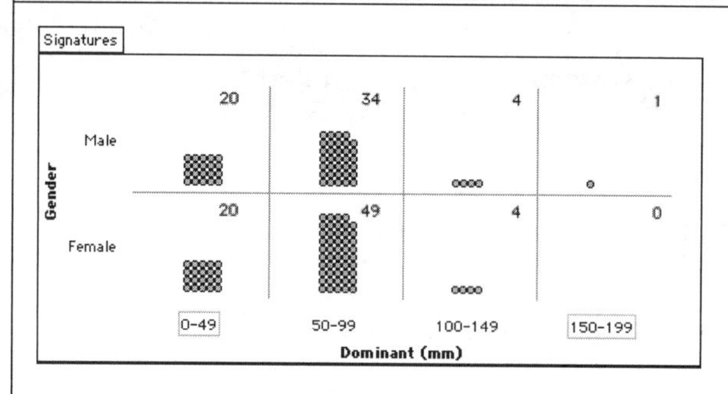	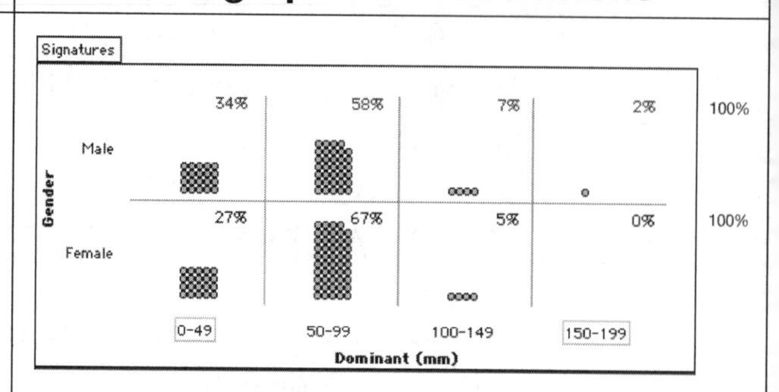
The same number of males (20) and females (20) have short signatures that are 0–49 mm long. 15 more females than males have medium length signatures that are 50–99 mm long.	A higher percentage of males (34%) than females (27%) have short signatures that are 0–49 mm long. 9% more females than males have medium length signatures that are 50–99 mm long.

You had to decide whether males or females have longer signatures. If females tend to have more letters in their names, then their signatures could be longer than males' signatures even if they wrote smaller.

4. You can use a formula to determine the average length per letter for each signature. For example, Marina Gray has these values:

   - Number of letters in name = 10
   - Length of dominant signature = 60 mm

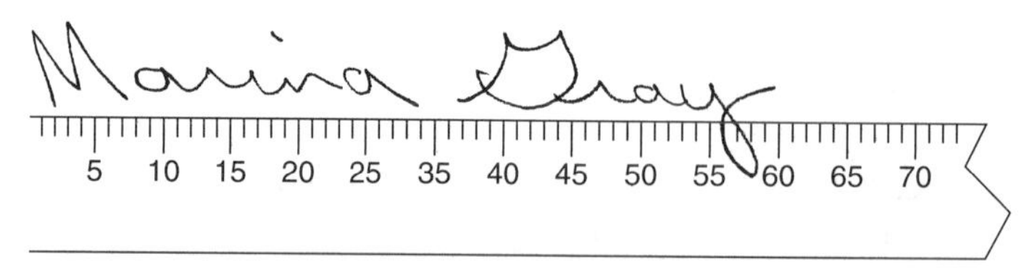

   a. What is the average letter length in Marina Gray's signature? How did you figure this out?

   b. What is the average letter length in your own dominant signature? ________

5. Use TinkerPlots to make a formula to determine the new attribute, *Letter_Length*, for all the students in the data set.

6. Make different plots such as line plots and box plots to compare the new attribute, *Letter_Length,* for the males and females. Keep notes in the table.

Similarities and differences between the two groups	What does this tell you about how the males' and females' signatures compare?

COMMUNICATE CONCLUSIONS

7. On a separate sheet of paper, write your conclusions for the question: How do the lengths of dominant signatures compare for this group of middle-school males and this group of middle-school females?

   Do your conclusions support your hypothesis? Why or why not?

   Make sure to follow the criteria on the Rubric for Making Comparisons.

OVERVIEW

In this lesson, the focus shifts from comparing two groups to investigating the relationship between two attributes. Students explore three different graphical representations of the two attributes. The lesson guides students in creating the graphs. Then, students progress to analyzing each graph to draw conclusions about the relationship between the two attributes. As the final step, students compare the representations. This lesson helps to prepare students for their work with scatter plots in Section 5.

Objectives

- Investigate whether there is a relationship between two attributes
- Explore ways to create plots to show the relationship between attributes
- Write high-quality conclusions by using evidence from the data

TinkerPlots

Class Time: One class period

Materials

- Grades and Lengths of Conclusions worksheet (one per student)
- Rubric for Investigating Relationships (transparency or one per student, on CD)

Data Set: Grades and Words.tp (data from 37 students; the teacher first graded their conclusions and then the students counted the number of words they had written.)

TinkerPlots Prerequisites: Students should be familiar with intermediate graphing.

TinkerPlots Skills: Changing bin widths and changing axis endpoints are explained in this lesson.

LESSON PLAN

Introduction

1. Introduce the new data set: a class of seventh-graders wrote conclusions about cats' weights and then counted the number of words in their conclusions. Students will be investigating the question: Is there a relationship between the number of words in students' conclusions and their grades? Do students who write longer conclusions tend to get

better grades? Point out that the seventh-graders counted the number of words in their conclusions after they had been graded—when they wrote the conclusions they didn't know that they would be looking at the length.

2. Before working with the data, have students write their hypotheses individually and then share them with the group (question 1). Samples are available in the answers.

Exploration

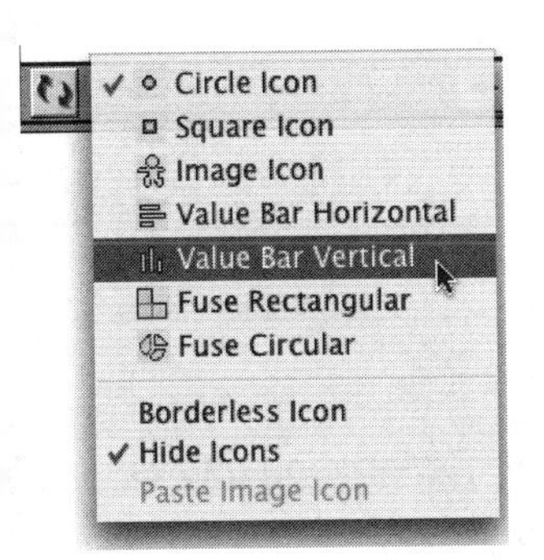

3. In questions 3–5, students use value bars and color to analyze the relationship between the two attributes. This is a new method for students, so you may want to demonstrate it. To make the graph, color by *Word_Count* and fully separate vertically. Then choose **Value Bar Vertical** from the **Icon Type** menu and click the **Order Horizontal** button. Finally, color by *Grade.* To help students see the relationship, ask these questions.

   - Do the conclusions with more words (the longer bars) have higher grades (darker colors)?
   - Do the conclusions with fewer words (the shorter bars) have lower grades (lighter colors)?

Wrap-Up

If time is short, have students make a list of their findings so that they can write their conclusions for homework.

4. Have students write their conclusions. Go over the Rubric for Investigating Relationships. Point out that this rubric is similar to the one that students have used before, but there are some differences. In previous lessons, students compared two groups, whereas in this lesson, they need to write conclusions about the relationship between two attributes.

5. Have a class discussion about the conclusions.

   - What did you find out about the relationship between word counts and grades in this data set? How can you tell from the data?
   - Which plot was most helpful for analyzing the relationship? Why?
   - What are some of the limitations of the data set? What suggestions do you have for ways to collect data that would help us further investigate the relationship between word counts and grades? [Limitations: This is a small sample of students, with grades from just

one assignment, and grades can be subjective. Suggestions: Collect data from a larger sample and more assignments, and have more than one teacher grade the assignment and then average the grades.]

- What other questions about word counts would you like to investigate? How would you collect the data? What do you think the data would show?

ANSWERS

Grades and Lengths of Conclusions

1. Sample student hypotheses:

   Longer conclusions tend to get higher grades than shorter conclusions.

   Jessica: Of the conclusions we read in class the longer ones tended to be better quality.

   Kiara: usually when something is longer it has more explanation and is sometimes clearer. It also gives more examples.

   D.J.: you can be more thorough and same [say] more in a longer conclusion.

   There is no relationship between the length of the conclusions and the grade.

   Monica: It depends on the content of the writing, not the length.

   Summer: Some can be short, precise and accurate or not. Some can be long and tedious or not.

   Rachel: you could write a short powerful conclusions or long but off-the-topic conclusion

   Su-Jung: if you miss your point, you get lower grade even if you have long essay.

5. Sample answer: The dark green bars are mostly at the right where the word count is highest, though there are some dark green bars in the middle. All the really light bars are at the left where the word count is lowest. So I would say that the short conclusions all got low grades, but the high grades were spread among the medium and long conclusions.

7. Sample answers:

   a. True. All the dots for the students who wrote 30–79 words are in the 71–80 and 81–90 bins. There are no dots in the 91–100 bin.

   b. Mostly true. The students who wrote 130–179 words have 8 students who got A grades, 2 students who got B grades, and no students who got C grades. This is better than the students who wrote 80–129 words: half of them got A grades and half B grades. None of the students who wrote 30–79 words got A grades: they are evenly split between the B and C grades. However, the one student who wrote 180–229 words also got an A grade, so maybe it would be better to say that students who wrote 130 words or more got the best grades.

9. Sample answer: This plot shows that the conclusions with fewer words tended to get lower grades than the conclusions with more words. Conclusions that got the lowest grades (71–80) had the lowest mean word count (53 words). The middle range of grades (81–90) had a higher mean of 93 words, and the highest grades (91–100) had a mean of 131 words. None of the conclusions that were graded 91–100 had fewer than 80 words, and all the conclusions that were graded 71–80 had fewer than 80 words. Some conclusions that were graded 81–90 had the same number of words as other conclusions that got lower grades or higher grades. As the grades go up, there is more variability in the word count, with the low grades having a range of 34 words, the medium grades having a range of 96 words, and the high grades having a range of 122 words.

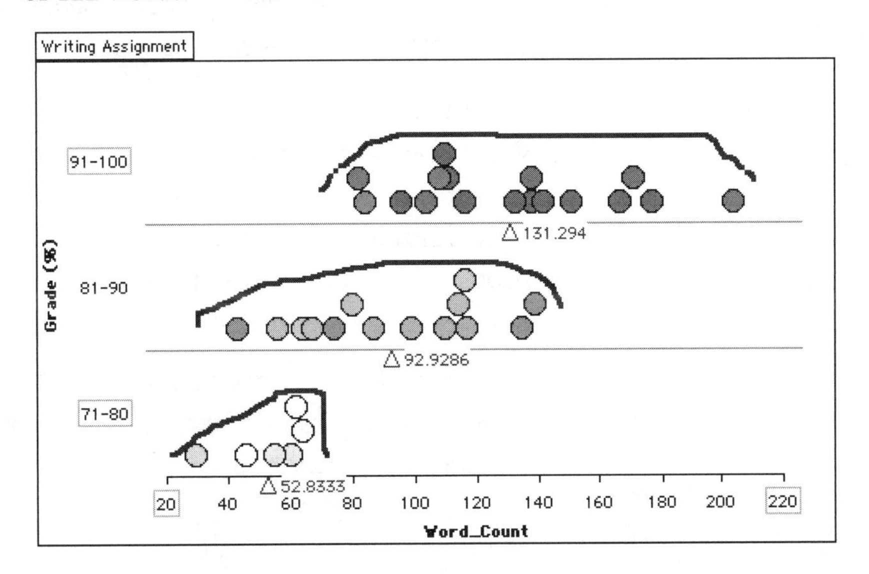

10. Answers will vary.

11. Sample student work:

Su-Jung: My conclusion is that many people with longer conclusion get high grade but length of the conclusion can be short and get high grade. Most people over 200 word-conclusion got 100%, bur [but] one person got about 95%. People who wrote less than 100 word-conclusions got less than 95% except for one person, who got 100% with 82 words. The scores of medium conclusions, ones with 82–114 words, were in between 76% to 100%, the scores of short conclusions, ones with less than 82 words, were less than 85%. Usually people with longer conclusions got higher grades, but some people got high grades with short conclusions.

Comments: The student describes the relationship between word counts and grades ("longer conclusions got higher grades"), and also points out exceptions to that conclusion. (She is using percentages for the grades.) One strength is that the student gives specific numbers for the groups. One weakness is that the student uses general terms such as "most" and "many" and does not gives specific percentages.

Summer: My conclusion is I think that quantity sometimes help quality, but the data shows no definite pattern. One person wrote 30 words and got a better score than someone who wrote 46 words. Someone wrote 151 words and got the same score as someone who wrote 146 words. A person who wrote 0-110 words got the same score as some people who wrote 110-120. There is no real pattern with the data, however more people who wrote more got good scores. Some people who didn't write as much got good scores, so there is no pattern with the data.

Comments: One weakness of the student's conclusions is that she focuses first on individual cases. This focus on individual exceptions seems to get in the way of the student being able to make a generalization about the group as a whole. She does write that more people who wrote more got good scores, but she does not give specific numbers to define what she means by "good scores" or "more people."

Alana: I conclude that the number of words does affect the grade. Most people with an average number have an average grade Most people with a very short paper have a low grade Most people with a very long paper have a high grade

Comments: The student seems to have a sense of what to look for to see a relationship between two attributes ("very short and low grade" and "very long and high grade"), but she doesn't use specific numbers. She does not explain what she means by a "very short" paper or a "very long" paper. It's also not clear what she means by "most."

D.J.: Is a longer conclusion better? Yes, I believe that in most cases, a longer conclusion is better than a short one. When you write more, you can explain more and tell things about the topic. Also, you'll be clearer than if you wrote a short, less detailed conclusion. A long conclusion is more complete, too, because everything to tell is told, instead of just part of it. Finally, more writing shows that you put more effort into the work, and the more effort, the better it will be. That is why I think that in most cases, a long conclusion is a better conclusion.

Comments: The student's conclusions are weak on all the rubric criteria, except that his writing is clear and easy to follow. He has written persuasively about his opinion but has not used the data as evidence for his conclusions.

Grades and Lengths of Conclusions

Name:

You will investigate the relationship between two attributes by using a variety of plots.

ASK A QUESTION AND MAKE A HYPOTHESIS

Is there a relationship between the number of words in students' conclusions and their grades? For example, do longer conclusions tend to get higher grades than shorter conclusions?

1. What do you think the data will show? Why?

ANALYZE DATA

2. Open the TinkerPlots file **Grades and Words.tp.** 37 seventh-graders wrote conclusions to compare the weights of male cats to female cats. After the teacher graded their conclusions and handed them back, students counted the number of words.

3. Use the value bars to make a bar graph of *Word_Count.* Order the bars so that the conclusions are in order from lowest to highest word counts.

4. Select the attribute G*rade* to color the bars by grade: the darker the color, the higher the grade.

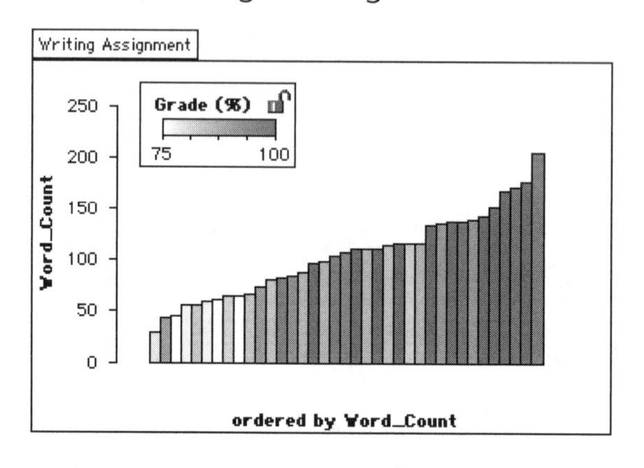

5. What did you find out about the relationship between the attributes from the bar graph? Do the bars for higher word counts tend to have higher grades (darker colors)? Do the bars for lower word counts tend to have lower grades (lighter colors)?

6. Make a grid of *Word_Count* and *Grade* like the one below. Change the intervals so that the grades start with 71 and have bin widths of 10 and the word counts start with 30 and have bin widths of 50. (To change the intervals, double-click inside the gray box.)

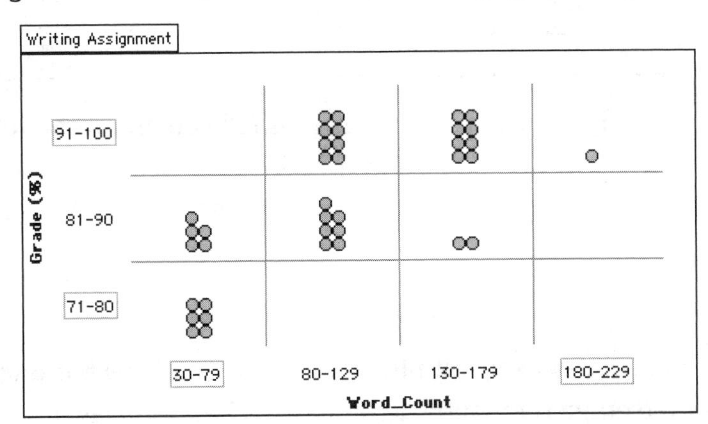

7. Is each statement correct? Give evidence from the grid.

   a. All the students who wrote 30–79 words got grades below 91.

   b. Students who wrote 130–179 words got the best grades.

8. Create a line plot for word counts. Then separate the data by grade into three groups: 71–80, 81–90, and 91–100, so that you have three line plots. (Double-click the boxes to change the intervals.) Find the mean word counts for each line plot and add them to this plot.

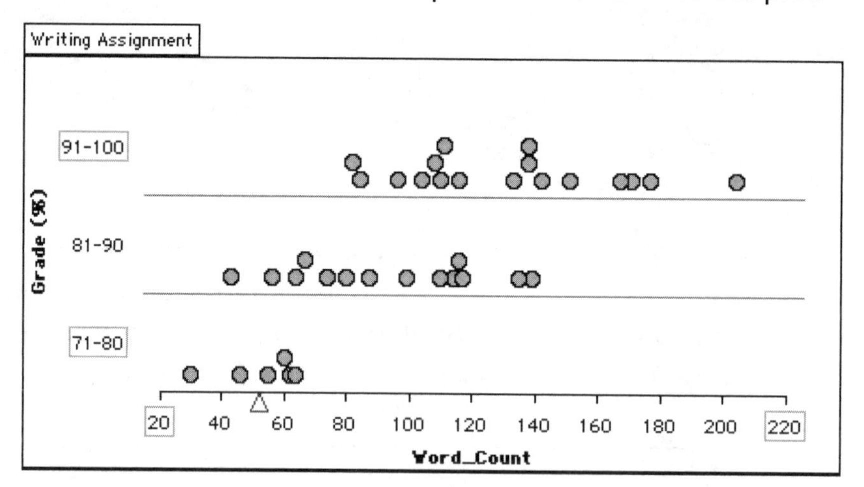

9. What do the line plots and means tell you about the relationship between word counts and grades?

10. You have worked with three different graphs: a bar graph with a color gradient, a grid, and line plots. Which graph was most helpful for analyzing the relationship between word count and grade? Why?

COMMUNICATE CONCLUSIONS

11. On a separate sheet of paper, write your conclusion for the question: Is there a relationship between the number of words in students' conclusions and their grades?

Section 4

Measures of Center and Histograms: Analyzing Safety Data

In this section, students investigate data about people who were injured while riding scooters and while using fireworks. They analyze and describe what's typical for these data sets, using visual and numeric comparisons. Students learn how to create and analyze histograms, which are useful for large data sets and ones with many different numeric values. They also work on improving their understanding of means, medians, modes, and their writing about data by addressing common misconceptions.

OBJECTIVES

Collecting Data

- Represent information from narrative descriptions on data cards (Lesson 4.1)

Analyzing Data

- Pose questions about data (Lesson 4.1)
- Determine what is typical for a group by examining several attributes (Lessons 4.2, 4.6)
- Create and interpret a variety of plots to analyze one or more attributes (Lessons 4.2, 4.6)
- Deepen understanding of measures of center: means, medians, and modes (Lessons 4.3, 4.4)
- Explore the effects of outliers on means and medians (Lesson 4.3)
- Address common misconceptions about means and medians (Lesson 4.4)
- Create, interpret, and compare histograms (Lesson 4.5–4.7)
- Compare histograms to other types of graphs (Lesson 4.5)
- Create, interpret, and compare circle graphs (Lesson 4.6)

Communicating about Data

- Create plots to communicate findings about a group of cases (Lessons 4.2, 4.6)
- Write summary statements about a group of cases (Lessons 4.2, 4.6, 4.7)
- Prepare a presentation using data as evidence for findings (Lesson 4.6)
- Write strong comparison statements that explain findings in the context of the data (Lesson 4.7)

Applications of Math Concepts from Other Strands

- Number and Operations: Use percentages to analyze data (Lessons 4.2, 4.6)

TINKERPLOTS SKILLS

In this section students use mostly intermediate graphing skills: making line plots, dividing data into equal intervals, and displaying percentages, means, and medians. They learn how to change values within a plot to watch the mean and median change, and how to hide cases. They also learn how to make histograms and circle graphs and change bin widths.

Introducing Data about Scooter Safety

OVERVIEW

The main goal of this lesson is for students to find key information in a written "injury report" and then record it on a data card. In previous lessons, the data had already been organized for students. This is the first time that they need to take a narrative description and translate it into values for a data card. Students tend to be engaged by the stories about the injuries, and this motivates them to investigate the data set in subsequent lessons.

Objectives

- Become familiar with the attributes in the data set and learn how the data were collected
- Take information from a written injury report and figure out how to represent it on a data card
- Make predictions about the data to increase interest in the actual data set

Offline

Class Time: One class period

Materials

- Scooter Injury Reports worksheet (one per student)

LESSON PLAN

Introduction

1. Introduce the context for the new data set, which consists of the kinds of injuries people got when they were hurt riding their unpowered scooters. These people went to the emergency room to be treated for their injuries. The Consumer Product Safety Commission (CPSC) collected data on injuries from 100 emergency rooms across the U.S. To find out more about the injuries, the CPSC did phone interviews with 28 people who were injured. Students will read injury reports from four phone interviews and then record that information on data cards.
2. To engage students with the new data set, you may want to have a brief discussion about sports-related injuries that students have sustained.

Exploration

3. Have students read each injury report and fill out a written data card to represent the information (questions 1–4). You may want to go through the first example with the whole class. Point out that the causes of the accidents are related to problems with the scooters, such as the brakes failing.

4. Ask students to make a prediction to answer the questions: Which of the four people do you think is most typical of the group of people who got scooter injuries? Why? (questions 5 and 6). Have a class discussion about the predictions. Point out that students will have the opportunity to see if their predictions are correct when they investigate the data set in subsequent lessons.

Wrap-Up

5. Ask students to generate their own questions about scooter injuries (question 7).
   - Which of these questions could you answer using this data set?
   - What other kinds of data would you need to collect to answer your questions?
   - How would you collect the data?

ANSWERS

Scooter Injury Reports

1. Female; 8 years; Arm and Hand; Hill; Brakes failed
2. Male; 10 years; Arm and Hand; Street; Handlebar problem
3. Male; 16 years; Leg and Foot; Driveway; Handlebar problem
4. Female; 41 years; Arm and Hand; Sidewalk; Handlebar problem

5., 6. Sample student work:

Jessica: 10-year-old male. Because males seem to be the ones who get the most scooter injuries. 10 years is also the age that many people get hurt at, and he had the same problem as most other people, a handlebar problem where he lost control. He also had the most common injury, hands and arms.

Comments: This student considered a variety of attributes in making a prediction.

Rachel: 8-year-old-female. It's because she's the youngest. Younger kids make more mistakes than older kids do and loosing balance is usual.

Comments: This student based her prediction primarily on age.

7. Sample student questions:

   How many kids in the US are hurt a year on scooters?

   What state has the most injuries?

   I would like to find out if they were alone or with friends.

   How many times have they been injured by scooter?

   What time of day were they injured?

   Whether girls or boys tend to use scooters more?

   I would like to know if people wear protective gear on their head, arms, and legs?

Scooter Injury Reports

Name:

Read each injury report and then fill in the values in the data card. Choose from these values.

Gender: Female Male

Injuries: Arm and Hand Leg and Foot Face and Neck Head Teeth Trunk

Location: Driveway Hill Home/Indoors Sidewalk Street Other

Cause: Brakes failed Handlebar problem Scooter broke Screw fell off Wheel fell off

1. An 8-year-old female was injured while riding a scooter on a hill. She was going too fast and could not stop her scooter. She said that the brakes failed, so she jumped off and fell. She fractured her right wrist.

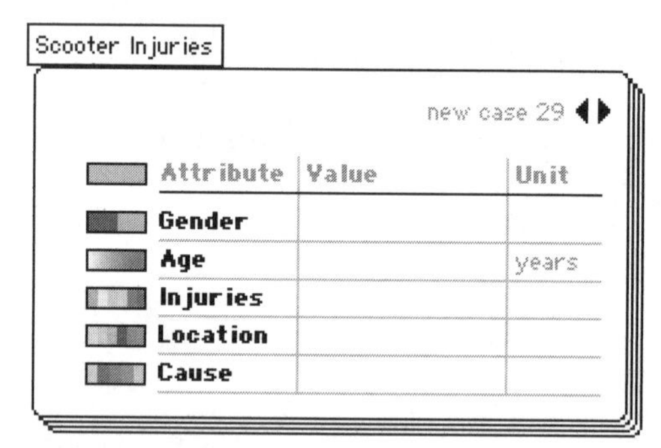

2. A 10-year-old male was injured while riding a scooter on a street in his neighborhood. He lost control of the scooter when the handlebars unlocked and slid down. He fell off and fractured his wrist.

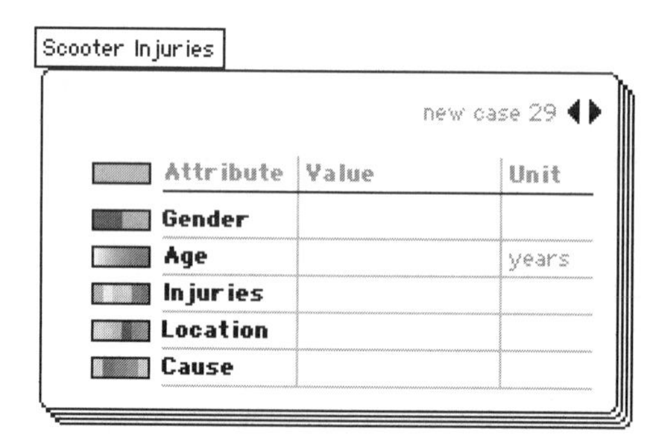

3. A 16-year-old male was injured in his driveway while riding a scooter. The handlebar became loose, causing him to lose control of the scooter and fall off. He sprained his ankle.

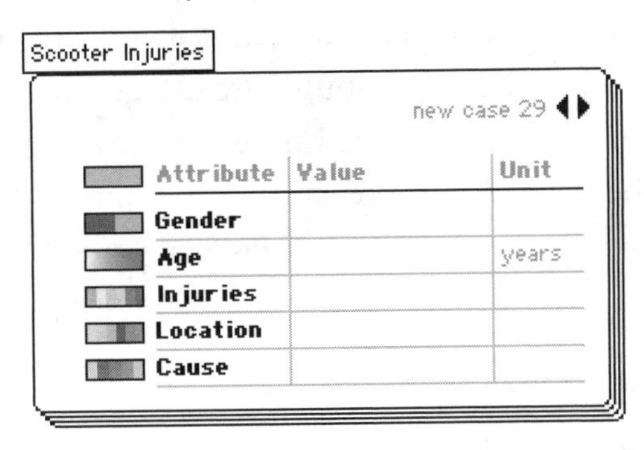

4. A 41-year-old female was injured while riding her son's scooter on the sidewalk. She lost control of the scooter because the handlebar became loose. She fell down and injured her hand and fingers.

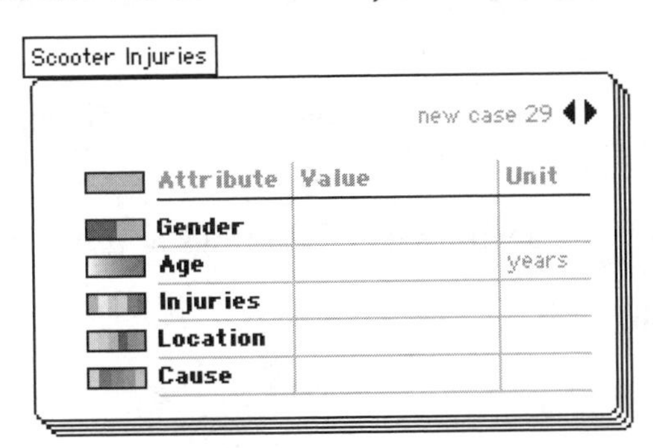

5. Think about the four people you read about and how they got injured. In the next lesson, you will be working with a data set of 28 people who got scooter injuries. Make a prediction about that data set: Which person do you think will be most typical of the group of 28 people with scooter injuries? Circle your choice.

   8-year-old female   10-year-old male   16-year-old male
   41-year-old female

6. Why do you think this person is typical of the group of 28 people with scooter injuries?

7. What would you like to find out about scooter injuries? Write two questions on your own paper.

Summarizing Data about a Group

OVERVIEW

This lesson is designed to help students recognize the importance of analyzing several attributes to determine what's typical for a group. Because four of the five attributes are categorical, students use counts, percentages, and modes for determining the common characteristics of scooter injuries. The context of preparing a presentation focuses students on how to best communicate their findings. They need to decide which plots to select and what points to make in their summary.

Objectives

- Use counts, percentages, and modes to analyze data
- Determine what is typical for a group
- Understand that it is helpful to examine multiple attributes to determine what is typical
- Create a variety of plots and decide which ones are best for communicating findings
- Write a summary of findings to use in a presentation

TinkerPlots

Class Time: One class period

Materials

- What Is Typical for Scooter Injuries? worksheet (one per student)
- Scooter Safety and Age worksheet (one per student, *optional*)
- Scooter Safety Extension worksheet (one per student, *optional*, on CD)

Data Set: Scooters.tp (data for 28 people with scooter injuries)

TinkerPlots Prerequisites: Students should be familiar with intermediate graphing.

LESSON PLAN

Introduction

1. Set the context by asking students to take on the role of the head doctor of an emergency room at a hospital. They have a lot of experience treating people with scooter injuries. The staff of the new emergency room wants to learn about scooter injuries. Students need to prepare a presentation answering these questions: Who typically gets injured? What are the most common scooter injuries?

2. Go over the steps in preparing for the presentation. The first step involves analyzing the data to determine what's typical for the group of scooter injuries. Students also need to select a case that they think is representative of the group so that they can include a story about that case in their talk. The second step involves creating three plots that they would show to communicate information about the injuries. The third step involves writing four summary statements that they plan to say in the presentation.

Exploration

3. Students use TinkerPlots to investigate the data to determine what is typical of the group. They create graphs for each of the five attributes and record their findings in the table for question 1.
4. Then they need to select a case that is representative of the group to use as an example in their talk (question 2). Here's one way to use TinkerPlots to find a representative case.
   a. Make graphs of all the attributes, and turn off the color (click the blue bar next to the Attribute column in the data cards).
   b. Highlight a case in the most common group in one graph, and look for where that case appears in other graphs.
5. Students create plots of the data to show in their presentation (question 3). Encourage students to create a variety of plots and then select the ones that they think would be best for communicating the information to other people.
6. Students write summary statements about the points that they would make in the talk (question 4).

Wrap-Up

7. Have a class discussion about students' approaches to analyzing the data. Point out that one of the limitations of the data set is that it contains only 28 cases. If students were preparing an actual presentation, it would be important to make generalizations based on a larger data set.
   - How did you figure out what was typical of the group of injured people?

- Is it possible to find means and medians for each of the attributes? Why or why not? [It's not possible to find means and medians for attributes with words as values: *Gender, Injuries, Location,* and *Cause.*]
- What kinds of plots did you choose for your talk? Why?
- What are the characteristics of high-quality plots for communicating information?
- What are the characteristics of high-quality summary statements?

8. Ask a few students to give their talks to the whole class. If time is short, have students write their summaries for homework and share their presentations in the next class.

Additional extensions using this larger data set are available on the *Digging into Data* CD.

9. As an offline extension, give students the Scooter Safety and Age worksheet, which has information about scooter injuries for a much larger data set from the Consumer Product Safety Commission. This gives students a fuller, more accurate picture of scooter injuries. Because these tables have only aggregate data, not data for individual cases, they give students a different perspective on data analysis.

ANSWERS

What Is Typical for Scooter Injuries?

1. Sample answer:

Attribute	What is typical for the group of scooter injuries?
Gender	**61% of the injured people are male and 39% are female, so being male is most typical.**
Age	**The middle 57% of injured people were between 9 and 12.5 years old. The median age for injured people is 10.5 years.**
Injuries	**The two most typical injuries are Arm and Hand (36% of cases) and Leg and Foot (32% of cases). The majority of people (68%) had one of these injuries.**
Location	**The two most typical locations are Street (31% of cases) and Sidewalk (27% of cases). The majority of people (58%) were injured in one of these locations.**
Cause	**The most typical cause was a Handlebar problem (50% of cases). Brakes failing was also quite common (36% of cases).**

2. Sample answers:

   a. Case 9

   b. I chose this case because he has the most typical values for all the attributes.

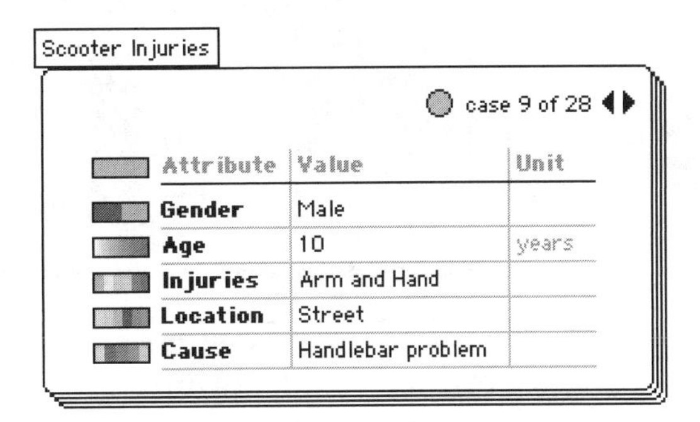

3. Sample answer: This graph shows the most typical ages of the people injured. There is a very tight center clump of 57% of values between 9 and 12.5 years, showing that these people get injured the most often.

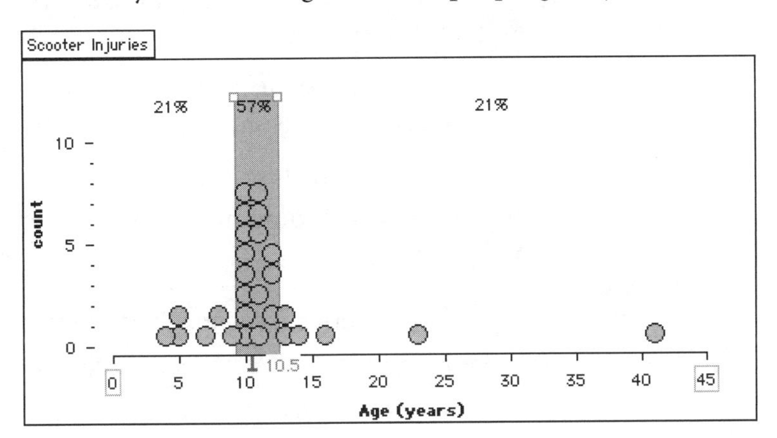

This plot shows that the two most typical injuries are Arm and Hand (36% of cases) and Leg and Foot (32% of cases).

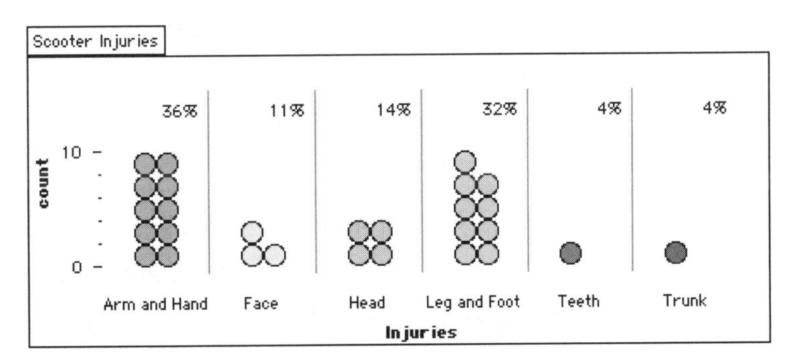

This plot shows that the most typical cause was a Handlebar problem (50% of cases). Brakes failing was also quite common (36% of cases).

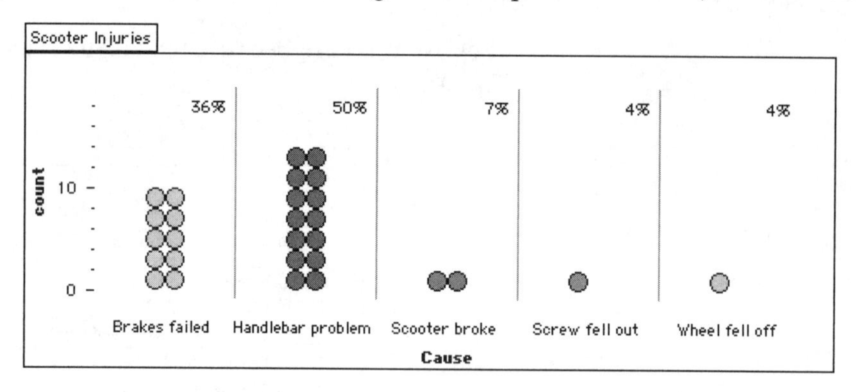

4. Sample student work:

Emily: More males have injuries than females. 17 vs 11, 6 diff Few people are over 15 years old Hand Arm and leg/foot injuries are common. Most common age for injuries is 10 years old.

Comments: This student wrote statements about three attributes: *Gender, Age,* and *Injuries.* The first statement provides specific numbers and a comparison ("17 vs 11, 6 diff"). The other statements are more vague; she uses the terms *few* and *common.* These statements would be improved by including more specific information, such as that 36% of the injuries were to arms and hands and 32% were to legs and feet.

D.J.: Handle bars causes the most injuries. The ratio of males to females for injuries is 17:11. Arms and hands are the body part of most injuries. 7–14 year olds take up 75% of the injuries.

Comments: These statements provide information on four attributes: *Cause, Gender, Injuries,* and *Age.* The student gave specific numeric information in the form of ratios and percentages. The student used the term *most* correctly in the first statement. In the third statement, however, arms and hands are the most common injury (36%) but not the majority of the injuries.

Thomas: I found out that males have more injuries than females. Most injuries are to the arm and hand. The location of the average accident is in the street. The most common cause was the handle bars.

Comments: This student included statements on four attributes: *Gender, Injuries, Location,* and *Cause.* One weakness is that the student did not provide any numbers or percentages to make the statements more specific. The student used the term *most* correctly in the fourth statement but incorrectly in the second statement. The student seems to have a misconception that *mode* means majority.

Scooter Safety and Age

1. a. 8–11 yr
   b. The most common way people aged 8–11 yr got injured was to fall after hitting something small. About 56% of the people in this age group (6,938 people), were injured this way. The next most common way was to fall while trying to stop. About 20% of people in this age group (2,492 people) were injured this way.
2. Sample answer: Actually, people in all the age groups tend to get injured for the same reason: falling after hitting something small. This table shows the percent of people in each age group who were inuried this way. This is the most common reason for all the categories except 16–19 yr. About 42% of people in that age group were injured by falling while trying to stop.

Age (yr)	0–3	4–7	8–11	12–15	16–19	20+
Percent of injured people	100%	61%	56%	68%	34%	45%

What Is Typical for Scooter Injuries?

Name:

Imagine that you run an emergency room at a hospital and have a lot of experience treating people who get injured riding unpowered scooters. You have been asked to talk to the staff of the new emergency room about scooter injuries. You need to prepare a short presentation on scooter injuries answering these questions.

- Who typically gets injured?
- What are the most common scooter injuries?

WHAT IS TYPICAL?

1. What usually happens in scooter injuries? Who tends to get injured, and how? Open the TinkerPlots file **Scooters.tp.** Use TinkerPlots to investigate all five attributes and keep notes in the table.

Attribute	What is typical for the group of scooter injuries?
Gender	
Age	
Injuries	
Location	
Cause	

2. To make the talk interesting, describe an example case of a scooter injury.

   a. Which case do you think is most representative of this group of injured people?

      Case ___________

   b. What are your reasons for choosing this case?

WHAT WOULD YOU SHOW AT THE TALK?

3. Prepare at least three plots to communicate your findings about the group of scooter injury cases to the staff. Experiment with different kinds of plots.

   - Make one plot of the attribute *Age*.
   - Choose the attributes for the other plots.
   - Make sure your plots are clear and easy to interpret.
   - Write an explanation of what each graph shows. (You can put your explanations in TinkerPlots using text boxes.)

Text

WHAT WOULD YOU SAY AT THE TALK?

4. On a separate sheet of paper, write at least four summary statements about the data that you would put in your talk. Make sure your statements answer the staff's questions and include specific numerical comparisons, such as percents.

Scooter Safety and Age

Name:

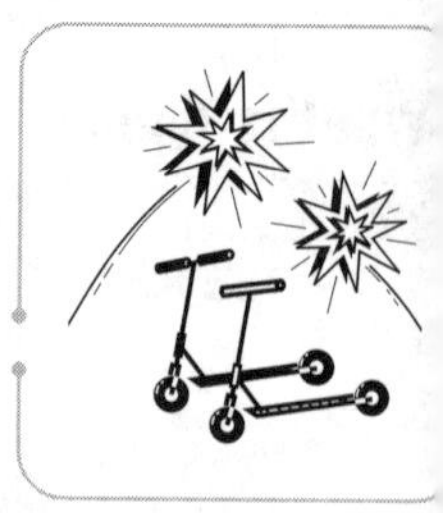

You've looked closely at data from 28 injured people. Now, you'll get a bigger picture by examining data from over 26,000 scooter injuries from the Consumer Product Safety Commission.

1. Analyze the data in the table to answer these questions.

   a. Which age group had the most injuries? ____________________

   b. What are the two most common reasons people in this age group fell and got injured? Give specific evidence from the table.

2. Do people in different age groups tend to get scooter injuries for different reasons? For example, are the most common reasons people aged 4–7 years get injured different from the most common reasons for people aged 12–15 years? Write your conclusions on a separate sheet of paper. Make sure to use evidence from the table.

	Age (yr)						
Reason for injury	**0–3**	**4–7**	**8–11**	**12–15**	**16–19**	**20+**	**Total**
Fell—Trying to stop	—	226	2,492	1,439	299	677	5,133
Fell—Scooter broke or failed	—	209	2,094	391	85	165	2,944
Fell—Hit something small	31	2,623	6,938	4,045	241	1,346	15,224
Fell—Turning	—	1,249	874	85	80	802	3,090
Total	31	4,307	12,398	5,960	705	2,990	26,391

Changing Means and Medians

OVERVIEW

This exploration is designed to build students' understanding of the differences between the mean and the median. Students learn that the median is less affected by extreme values than the mean. For this reason, the median is a good choice for describing what's typical for data sets that have extreme values. The mean tends to be the first choice for students, perhaps because they are more familiar with it (through grades, sports, etc.). This lesson helps students to consider when the median might be a better choice.

Objectives

- Learn about the concept of an outlier in an informal way
- Explore the effects of an outlier on the mean and median
- Deepen understanding of differences between the mean and the median, particularly that the median is less affected by extreme values than the mean

TinkerPlots

Class Time: One class period

Materials

- How Do Outliers Affect Means and Medians? worksheet (one per student)

Data Set: Scooters.tp (data for 28 people with scooter injuries)

TinkerPlots Prerequisites: Students should be familiar with intermediate graphing.

TinkerPlots Skills: Hiding cases and changing case values are explained in this lesson.

LESSON PLAN

Introduction

1. Introduce the task by connecting to the previous lessons. When your students looked at the ages of the people with scooter injuries, they probably made comments about the 41-year-old woman because she is so much older than the other people in the data set. Explain that this case is called an *outlier* because the value for age is unusually large in comparison to the other values.

Note: At the middle-school level students work with an informal definition of outlier: a value that is unusually large or small in comparison to the other values in the data set. Statisticians use formal methods for determining outliers. Using one definition, a value is considered an outlier if the distance from the value to the first quartile or third quartile (whichever is nearer) is more than 1.5 times the interquartile range.

2. Introduce the question that students need to investigate: How does an outlier affect the mean and the median? Explain that students will have the opportunity to change the age of the 41-year-old to see what happens to these measures. Ask the class to come up with the lowest and highest reasonable ages for someone riding a scooter.

Exploration

3. Have students work on the task independently or in pairs. They begin by finding the mean and median ages for the data set with all 28 values (question 2). Then, they see what happens to these measures when they remove the data for the 41-year-old (question 4).

   Troubleshooting tips for hiding the case of the 41-year old:

   - Note that you will hide the case only in the plot that is selected.
   - When you hide a case, it is not included in the calculation of any measures in that plot.
   - To hide multiple cases at once, draw a selection rectangle around them.

4. Students tend to get very engaged with the task of changing the age to see the effect on the mean and the median. They can try two different methods to change the age of the 41-year-old: typing a new age on the data card and dragging the value in the plot. Directions for each method are provided on the worksheet. You may also want to do a whole-class demonstration.

   Troubleshooting tips for dragging values:

   - To get the effect of changing the mean and median as you drag values, it's important that the scale of the plot be locked. Double-click an endpoint and click **OK** to lock that endpoint. You can change the endpoint value, but you don't have to. You do have to lock each endpoint separately.

- Note that you have to choose the **Drag Value** tool again every time you let go of the mouse. If you don't, dragging will rescale the plot as usual. To keep the **Drag Value** tool selected, hold down the Alt (Win) Option (Mac) key.
- If you drag a case off the axis, it will move into the stack of excluded cases and the axis will show that there's more data off the end.

5. As you circulate, ask students, "How can you get the median to change by changing the age of the 41-year-old?" [To change the median, you need to change the age to a value that is less than the original median of 10.5 years.] The lowest the median will go is 10 years old. Students tend to be surprised that the median doesn't change when they increase the age.

To explore this further, ask students to see if they can increase the median by adding a new case to the data (the data set will now have 29 cases). By adding a new case that is higher than the original median of 10.5, the median will increase. Show students how to use the **Add Case** tool and ask them to investigate: What new age could you add to the data set to increase the median? (Choose the **Add Case** tool and click in the plot over the age that you want. A new case will be added with all the other values blank.) Another way to increase the median is to change one of the ages that is below the median to an age above the median. Ask students to explore the question: What happens to the median if you change the values of other cases besides the 41-year-old?

Wrap-Up

6. Have a class discussion to help students clarify the differences between means and medians, particularly that medians are less affected by extreme values than means.

- Why did the mean change more than the median when you removed the 41-year-old from the data set?
- What was the highest mean you could get by changing this one value? The lowest?
- What was the lowest median you could get by changing this one value?

- Why doesn't the median change when you increase this one age? What would you need to do to the data set to increase the median?
- What did you learn about means and medians from this activity?
- Which measure—mean or median—do you think is more representative of the group of ages in this data set? Why?

ANSWERS

How Do Outliers Affect Means and Medians?

2.–4. Answers for question 3 will vary.

	Mean (yr)	Median (yr)
Data set with all 28 cases	11.75	10.5
Data set without the 41-year-old	10.67	10
How much did the measure change?	1.08	0.5

5. Sample student work:

   Jessica: Because outliers don't effect the median as much as they do the mean. So, when you take away the outlier the median doesn't have much to change while the mean does.

   Eana: because the mean is the average, so with one less that is 41 yrs old it goes much lower. But for the median its not much because it's the middle #,

   Su Jung: Because 41 raised the total a lot and when it was gone it showed the true mean of the data.

7. Answers will vary.

9. Students should try ages in all parts of their chosen range (of youngest and oldest ages for riding scooters), but not beyond it. For example, changing the age to 30 yr gives a mean of 11.3 yr and a median of 10.5 yr. Changing the age to 5 yr gives a mean of 10.3 yr and a median of 10 yr.

10. a. Answers will vary depending on the student's choice for lowest age. A lowest age of 3 yr gives a mean of 10.39 yr.

    b. 10 yr; any age below the original median of 10.5 yr will give this value.

11. a. Answers will vary depending on the student's choice for highest age. A highest age of 60 yr gives a mean of 12.43 yr.
    b. 10.5 yr; any age above the original median of 10.5 yr will give this value.
12. Any age above the original median of 10.5 yr will raise the median. If the age is between 10.5 and 11 yr, the new median age will be the same as the age added. If the age is above 11 yr, the new median age will be 11 yr.
13. Sample student work:

    Jessica: I learned that means can be greatly altered by an outlier while medians are not.

    Justin: Medians don't change much, but means change drastically.

    Eana: That each age changes the mean, the higher the age, the higher the mean, but for the median most of the time it stays the same.

    Ian: The median depends on the # of #s in a data set while the mean depends on the size of the #s.

14. Sample student work:

    Jessica: Median: because even though the mean is the average, it can be thrown off by outliers, while the median (the middle number) is not.

    Ian: Median because with the mean the data could be off because of a few odd balls.

    Comments: These students recognize that the median is a good choice for representing a data set that includes outliers.

    Rand: Mean because it sums up all of the ages, not just the middle one.

    Comments: This student seems to have a misconception that the median is not representative of the group.

    Nicole: Mean. I chose mean because it is more accurate. The median is just the middle number, and is usually the same as the mode.

    Comments: This student seems to share the misconception with the previous student that the median doesn't represent the group. She has an additional misconception that the median is usually the same as the mode.

Justin: Mean. Median doesn't tell much. The median is judged by what's above and below not that number.

Comments: This student seems to have a similar misconception about the median.

Yom: Mean because mean changes if one number is change but median doesn't. It is more specific than median.

Comments: The student has chosen the mean because she thinks it is more "specific." She is not focusing on how representative the mean is of the data set.

How Do Outliers Affect Means and Medians?

Name:

When you look at the ages of the people who got scooter injuries, one person stands out from the rest: the 41-year-old woman. This case is called an *outlier* because its value for age is unusually high in comparison to the other age values in the data set. (A case with an unusually low age value would also be an outlier.) In this lesson you'll explore how this outlier affects the mean and median ages for the data set.

1. Open the TinkerPlots file **Scooters.tp**. Make a line plot of the ages. Lock the scale by double-clicking the gray box at the right end of the axis and clicking **OK.**

2. Find the mean and median ages for people who got scooter injuries. Fill in the first row of the table below. (To find the values, click the arrow to the right of the mean button and choose **Show Numeric Value(s).**)

3. What do you think will happen to the mean and median if you remove the data for the 41-year-old? Circle your choices.

   The mean will: Stay the same Go up Go down

   The median will: Stay the same Go up Go down

4. Hide the data for the 41-year-old: Click on the dot for the 41-year-old. Go to the **Plot** menu and choose **Hide Selected Cases.** Fill in the rest of the table. Round to the nearest hundredth.

	Mean (yr)	Median (yr)
Data set with all 28 cases		
Data set without the 41-year-old		
How much did the measure change?		

5. Why did the mean change more than the median?

6. Put back the data for the 41-year-old: Go to the **Plot** menu and choose **Show Hidden Cases.** Then mix up the data and make a new line plot of *Age.*

7. What is the youngest age you think someone would be riding a scooter? __________

   What is the oldest age you think someone would be riding a scooter?
   ____

8. Change the scale to go from the youngest to the oldest ages you chose: Double-click the box at the left end of the scale. Enter the youngest age for **Axis starts at** and click **OK.** Repeat at the right end for the oldest age.

9. Change the age of the 41-year-old case to see what happens to the mean and median. There are two ways to change the age.

   - Click on the dot for the 41-year-old to show her data card. Click on the value 41 and type in a new age.
   - Choose the **Drag Value** tool. Click on the dot for the 41-year-old. Move the dot along the scale to see how the values for mean and median change. The age will change on the data card.

   Keep your notes in the table (round to the nearest hundredth).

New ages I tried (yr)	New mean (yr)	New median (yr)

10. What is the lowest mean and the lowest median you can get by changing this one age? (Do not use an age below the youngest age you chose in question 7.)

    a. Lowest mean: __________ What age did you use? __________

    b. Lowest median: __________ What age did you use? __________

11. What is the highest mean and the highest median you can get by changing this one age? (Do not use an age above the oldest age you chose in question 7.)

    a. Highest mean: ________ What age did you use? ________

    b. Highest median: ________ What age did you use? ________

12. You can also change the mean or the median by adding a case to the data set. Pick an age that will raise the median. Add a case with that age: Choose the **Add Case** tool and click in the plot over the age that you want. A new case will be added.

    What age did you add to raise the median? Why? How much did the median change?

13. What did you learn about means and medians from this activity?

14. Which measure—mean or median—do you think is more useful for summarizing the data about the ages of this group of people who got scooter injuries? Why?

Understanding Means, Medians, and Modes

OVERVIEW

The goal of this lesson is to directly address some common misconceptions and mistakes that students make when they interpret and determine means, medians, and modes. They need to work backwards from a measure, such as a mean of 85, to create a data set that makes the conclusion incorrect. Through the process of figuring out and explaining why a conclusion could be incorrect, students clarify their own understandings of these measures.

Objectives

- Address common misconceptions about means, medians, and modes
- Apply understanding of means, medians, and modes to create example data sets that have specific measures

Offline

Class Time: One class period

Materials

- Jumping to Conclusions worksheet (one per student)
- Means, Medians, and Modes worksheet (one per student, on CD)
- Jumping to Conclusions transparency (*optional,* on CD)

LESSON PLAN

Introduction

1. Start with a discussion: "What does it mean to jump to conclusions?"
2. Introduce the goals and tasks. Go through an example with the class to show how to create an example data set that would make the conclusion incorrect. You may want to use the Jumping to Conclusions transparency.

Exploration

3. Have students work in pairs to create data sets for the other questions and to write explanations of why the conclusions could be incorrect.

4. For additional challenges, ask students to
   - Create data sets that have more values.
   - Create data sets that make the conclusions correct.
5. Have a class discussion about the reasons why each conclusion could be incorrect. Ask a few students to share the data sets they created and the strategies they used to create them.
   - Why could the conclusion be incorrect?
   - How did you come up with a data set that makes the conclusion incorrect?

   If some students worked on the additional challenges, have them share their strategies.
6. Have students work individually or in pairs on questions 5 and 6. Point out that the focus of these questions is on common errors that people make when they find means and medians. Students need to find the errors and explain why the methods are incorrect.

Wrap-Up

7. Have a class discussion about methods for finding means and medians.
   - Why are Rose's and Micah's methods for finding medians incorrect?
   - What other kinds of mistakes might students make in finding means and medians? What are ways to avoid those mistakes?
8. To close, have students complete the Means, Medians, and Modes worksheet from the *Digging into Data* CD.

ANSWERS

Jumping to Conclusions

1. A data set that makes Monica's conclusion incorrect is: 80, 80, 80, 80, 90, 90, 90, 90. Here, the mean is 85 but no one in the class scored an 85. Sample student work:

   Ian: *The data set might not even have a 85 in it. The mean is the average not the mode.*

   Emily: *Because the mean doesn't have to be part of the data set.*

2. Leo is confusing median with mean. Leo would be correct if the PTO had raised a *mean* of $1000. The mean is the sum of the values, divided by the number of values, which in this case is 5. So you can work backwards from the mean to figure out the total. The median divides the data set into two halves when the values are in order from lowest to highest. About 50% of the data are below the median and about 50% are above it. To result in a median of $1000 the PTO could have raised two amounts below $1000, one amount of $1000, and two amounts above $1000. Unlike the mean, the median does not provide information about the sum of the data. A data set that makes the conclusion incorrect is: $800, $900, $1000, $3000, $5000. The median is $1000, and the total amount raised is $10,700 (not $5000). Sample student work:

   Emily: Median is middle no. not average. One of the amounts of money could be $500,000.

3. The mode is the response that occurs most frequently. In this situation, the students have a choice of four beverages. If an equal number of students selected each beverage, 25% of the students would have selected each one. There would be no mode in that case. If one more selected chocolate milk than the other three choices, chocolate milk would be the mode. However, only slightly more than 25% would have selected it. Mode means most frequent response, but does not necessarily mean that most (more than 50%) of the students made that response. A data set that shows how the conclusion could be incorrect is: whole milk—25; skim milk—20; chocolate milk—30; orange juice—25. In this case, 30 students selected chocolate milk, so it is the mode. However, 70% of the students did not select this choice. Sample student work:

   Kiara: It could be incorrect because it could only be the mode by a few.

   D.J.: The mode could be less than 50%

4. The mean is the sum of the values divided by the number of values. The mean is affected by extreme values. If one or more students jump rope a large number of times, this will increase the mean. Additionally, the mean is a computed value and it may or may not equal a value in the data set. If 75% of the students jumped fewer than 100 times, the mean could still be over 100. A data set that shows how the conclusion could

be incorrect is: 81, 89, 92, 93, 95, 98, 178, 202. The mean is 116. Sample student work:

Kiara: There could be 1 outlier that caused all the #s to go up

D.J.: A mean doesn't have to be someone's score and a person with many jumps could bring it up

5. The mean is the sum of the values divided by the number of values. In this case, 0 is a value for 10 of the 24 students. Therefore, leaving out the zeroes would result in omitting data for almost half the students. Consequently, the mean would not be representative of the class and would give the impression that students are absent for more days than is typical. Example data set of days absent: 0, 0, 0, 0, 0, 0, 0, 0, 1, 1, 1, 1, 2, 2, 2, 2, 3, 3, 5, 9.

   Mean calculated by including the zeroes: $\frac{32}{20} = 1.6$ days

   Mean calculated by excluding the zeroes: $\frac{32}{12} = 2.7$ days

   Sample student work:

   Rachel: It's incorrect because those students still exist. When you divide the total, you have to include them.

   Emily: If you leave out people who are never absent all the data you will get will be over 0 and the mean will be wrong and too high.

6. Neither Rose nor Micah is correct. The median divides the data into two halves. About 50% of the data is below the median and about 50% is above it. To find the median, Rose and Micah must first put the data set in order from the lowest value to the highest value: 17, 19, 20, 21, 22, 22, 23. The median is 21. Another common error about the median is to omit repeated values, for example the two 22s, when putting the values in order. Sample student work:

   Jessica: Rose is incorrect because it doesn't matter when you scored the points, it matters where they lie in the data. Micah is wrong because the median is not just the number in between the highest and the lowest numbers. It is the number in the middle of all the numbers in the data.

Jumping to Conclusions

Name:

Sometimes when people hear a statistic, such as "The mean is four hours," they jump to conclusions that may be incorrect. In questions 1–4, each conclusion *could* be incorrect depending on the values in the data set. Explain why it could be incorrect and make up an example data set that illustrates how the conclusion could be incorrect.

1. The teacher said, "The mean score on the math tests was 85."

   Monica said, "Most of the students in the class got an 85."

   Example scores: 70, 70, 70, 85, 90, 95, 100

   This data set has a mean of 85, but 85 is not the most common score.

   Your scores:

2. The principal said, "This year, our school had 5 fundraising activities to raise money. The median amount raised was $1000."

   Leo concluded, "The school raised a total of $5000 this year from the fundraising activities."

3. Students can choose from 4 beverages at the school cafeteria: whole milk, skim milk, chocolate milk, and orange juice. The cafeteria director said, "Chocolate milk is very popular. It is the mode."

   Jake concluded, "More than 50% of the students get chocolate milk."

4. During gym period, 8 students participated in a jumping rope contest. They recorded the number of jumps for each student. The gym teacher said, "The mean is 120 jumps per student."

   Cindy said, "I don't think that the mean is correct because 50% of the students did less than 100 jumps."

   Siena said, "The mean has got to be wrong because no one jumped exactly 120 times."

5. Tanya and Chris wanted to find out, "What is the mean number of days students in our class were absent last month?" Here are the numbers of days absent for each student.

0	0	0	0	0	0	0	0	1	1	1	1	2	2	2	2	3	3	5	9

   Tanya said, "There are 20 students in the class. Eight students were absent 0 days. They were here every day."

   Chris said, "I don't think we need to include the zeroes when we calculate the mean. It won't make a difference if we leave them out."

   Why is Chris's method incorrect for finding the mean?

6. Rose played in 7 basketball games. Here are the points she scored.

Game	#1	#2	#3	#4	#5	#6	#7
Score	17	21	22	19	20	23	22

   Rose said, "My median score is 19 because that's the score I got during the middle game."

   Micah said, "I think your median score is 20 because 20 is halfway between your highest and lowest scores: 17 and 23."

   Why are Rose's and Micah's methods incorrect for finding medians?

Representing Data with Histograms

OVERVIEW

This lesson focuses on a new graphical representation, the histogram. Histograms are often more challenging for students to understand than bar graphs because the data are grouped into intervals on a continuous scale, and the bars touch. Students go through the process of creating a histogram by hand to learn how to group the data into equal intervals called bins. This will help them to understand the histograms that they create with TinkerPlots in the next lesson.

Objectives

- Learn how data are represented in histograms
- Create histograms by grouping data into bins and then drawing the graph
- Compare histograms to other kinds of graphs
- Interpret histograms and summarize findings

Offline

Class Time: One to two class periods. *Note:* If your students are already familiar with histograms, you may want to skip the Histograms worksheet

Materials

- Histograms worksheet (one per student)
- Create Histograms by Hand worksheet (one per student)
- Measure Your Signatures worksheet from Lesson 3.1 (one per student, *optional*)
- Ages for Scooter Injuries transparency (on CD)
- Compare Three Graphs transparency (on CD)
- Fix the Histogram transparency (*optional,* on CD)
- Graph paper
- Tape measures or millimeters rulers (*optional*)

LESSON PLAN

Introduction

1. Introduce the goal of learning about a new graph, histograms. Histograms are particularly useful for representing large data sets. In the next lesson, students will be working with a large data set about injuries that were caused by fireworks.

2. Go over the basic characteristics of histograms using the Histograms worksheet. Display the Ages for Scooter Injuries transparency and point out the different parts of a histogram. Explain how to read the scale. For example, the 8–12 bin includes injured children ages 8, 9, 10, and 11 years but not 12 years. The third bin contains the 12-year-olds. Ask students questions to help them learn how to read the histogram.
   - Which bin has data for injured 5-year-olds? 8-year-olds?
   - Which bin has the most values? [8–12]
   - How many bins have no values? [4 bins]
   - How many values are in the second bin (4–8)? [4]
   - How would you describe the shape of the distribution? What does that tell you about the ages of the injured people?
3. Have students work through questions 1 and 2 on their own. Go over the answers by having students add cases to the histogram on the transparency.
   - How many bins do you need to add to include the case of the 50-year-old? What numbers do you need to add to the scale? [You need to add two bins: 44–48, 48–52.]
4. Introduce the context of the next part of the activity: the weights of students' backpacks and the connection between heavy backpacks and some back problems. Before students look at the graphs, ask them to make a prediction: What do you think middle-school students' backpacks typically weigh? After discussing some estimates, introduce the three graphs and point out that they represent the same data set of 41 fifth- and seventh-grade students.

You may want to use the Compare Three Graphs transparency.

5. Ask students to compare the graphs to answer questions 3 and 4. Discuss the answers.
   - Where in each graph are the data for the backpacks that weigh 19 lb?
   - How many students had backpacks that weighed 19 lb? Which graph or graphs can you use to get that information?
   - On the line plot, 19 is the mode, but on the histogram it's not in the tallest bin. Why?
   - What points would you make to summarize the data?
   - Which graph would you choose for your article? Why?

Exploration

6. Introduce the Create Histograms by Hand activity. You can use the data set provided with the lengths of signatures or have students create their own data. Histograms are a good representation for measurement data because there are often so many different values that it makes sense to group them. Here are some data collection ideas.
   - Use the data that students collected on their signatures in Lesson 3.1. If your students did not do that lesson, you can use the Measure Your Signatures worksheet now.
   - Have students measure the distance around their right wrist in millimeters. If you don't have tape measures available, students can wrap a strip of paper around their wrist and then measure the length.
7. Demonstrate the steps for making a histogram to the whole class. Then, have students work in pairs to create their own histograms. Students can choose different bin sizes, so they will create a variety of histograms to compare.

If some students are having difficulty, suggest that they use the histogram with bin width 10 (question 4) to help them make a histogram with bin width 20.

Wrap-Up

8. Display the Fix the Histogram transparency. Point out that the bin width is 12 and that the histogram represents the same signatures data (from question 1) that students used for drawing their histograms. Ask students to find the errors by using the information in their table (question 2) and the histograms they made and then figure out how to correct the errors.

   These are the errors.
   - Second bin has too many values in it (12 instead of 8)—the bin is too tall.
   - Fourth bin shouldn't have a value—it should be a gap.
   - Fifth bin width should be 12, not 13. It should have the interval 78–90, not 78–91.
   - Sixth bin width should be 12, not 14. It should have the interval 90–102, not 91–105.
   - Seventh and eighth bins have correct bin widths, but have incorrect intervals due to other errors. Seventh bin should have the interval 102–114 and eighth bin should have the interval 114–126.

Skip this activity if students worked with their own data. They need to be familiar with the signatures data set (on the Create Histograms by Hand worksheet) to find the errors.

Here is a correct histogram with bin width 12.

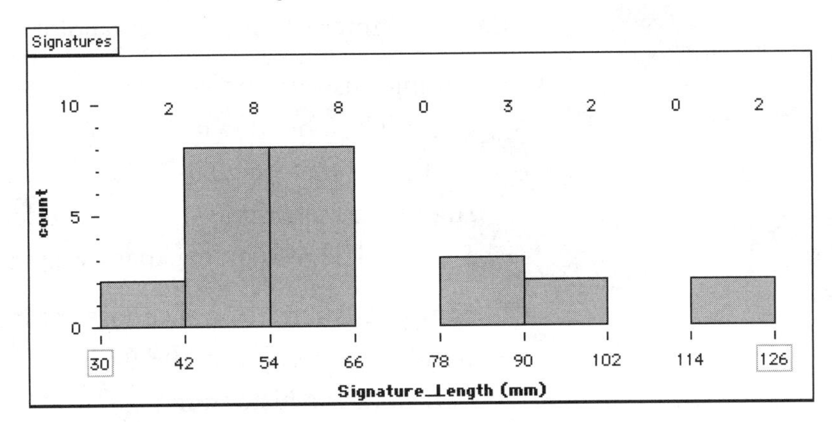

9. Have a class discussion to summarize what students have learned about histograms.
   - Why is it important for all the bins in a histogram to be the same width or size?
   - What are some things to remember when making histograms?

ANSWERS

Histograms

1. The 16-year-olds are in the fifth bin (16–20).
2. Here is the completed histogram.

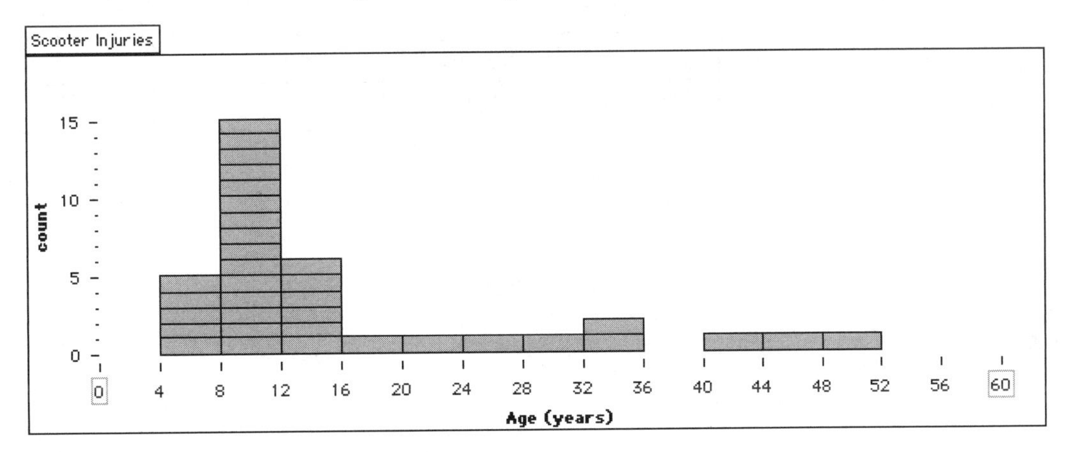

3. The data for the 19-lb backpacks is above the 19 on the line plot and in the 15–20 interval for the interval plot and histogram.

4. a. Sample answer: There is a large center clump between 5 and 25 lb that slopes down to the right, so most students (91%) carry backpacks that weigh 5–25 lb. The lighter backpack weights are slightly more common: 51% of students carry backpacks that weigh 5–15 lb. The median backpack weight is 14 lb.

   b. Sample answer: I would choose the interval plot. You can see the trend of the weights better than in the line plot, and I think it's easier to read than a histogram.

Create Histograms by Hand

1.

31	40	43	43	45	45	47	50	50	52	59	60	60
60	60	60	65	65	80	80	83	92	95	115	116	

2.

Bin	Values	Number of values
50–59	**50, 50, 52, 59**	**4**
60–69	**60, 60, 60, 60, 60, 65, 65**	**7**
70–79		**0**
80–89	**80, 80, 83**	**3**
90–99	**92, 95**	**2**
100–109		**0**
110–119	**115, 116**	**2**

3. a. 119

   b. 7

4. Bin width: 10

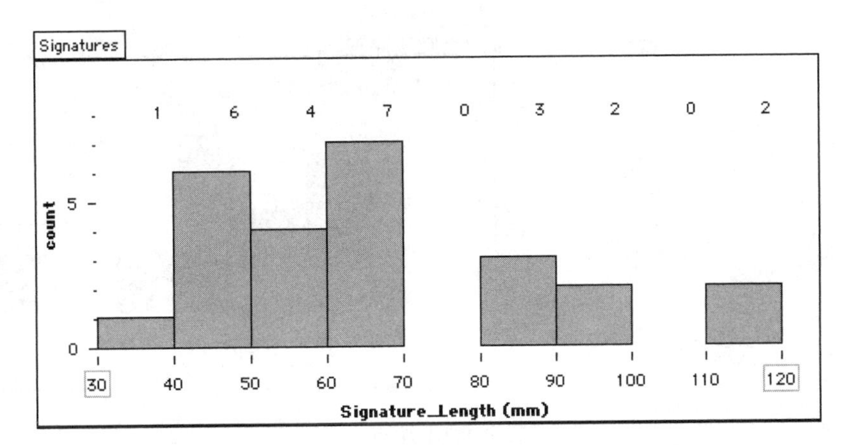

5. b. If the bin width students chose in 5a is less than 10, there will be more bins in this histogram, and if it is greater than 10 there will be fewer bins. The bins may or may not be taller. It depends on where in the intervals the values are. If there are very few bins, they will be taller.

7. Here are histograms for the various bin widths.

Bin width: 6

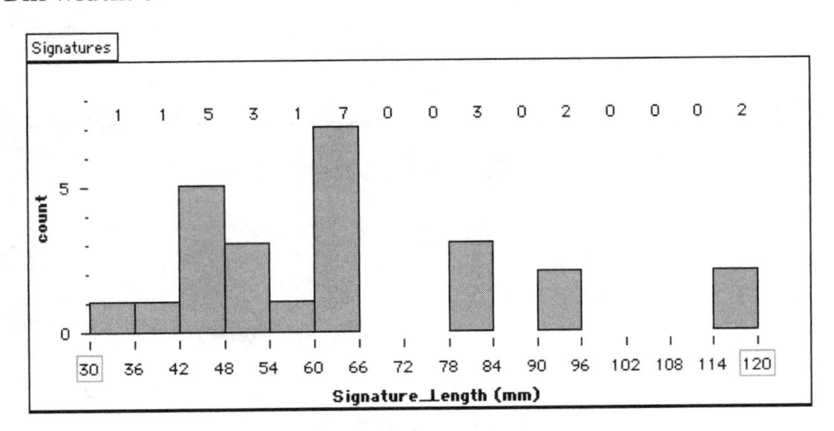

Bin width: 7

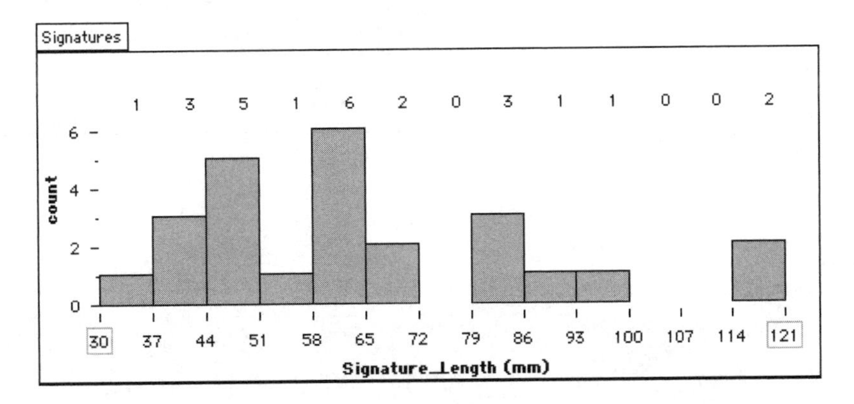

Bin width: 8

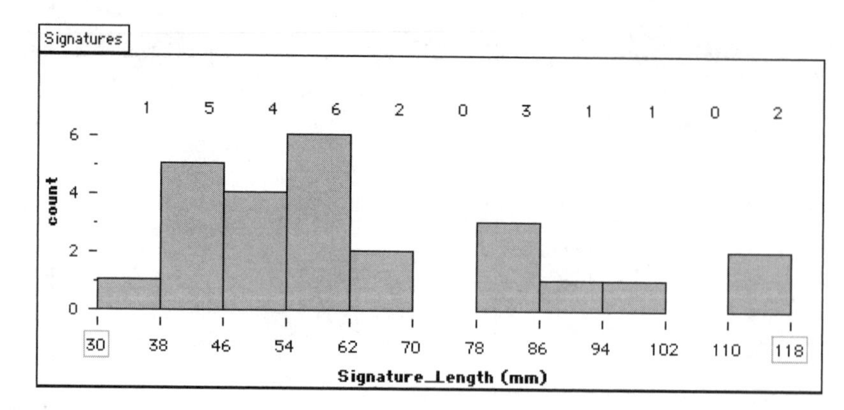

Bin width: 9

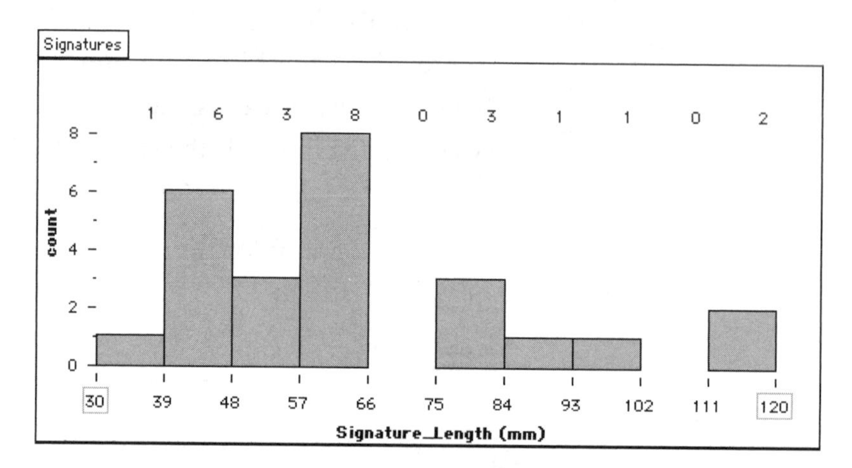

Bin width: 15

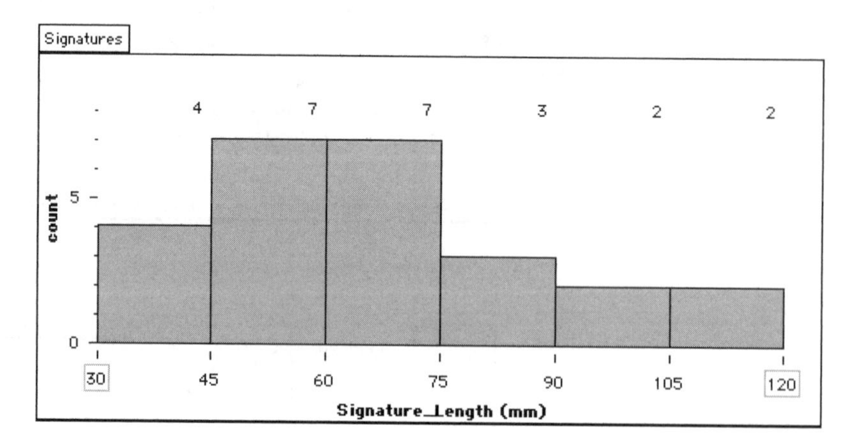

Bin width: 20

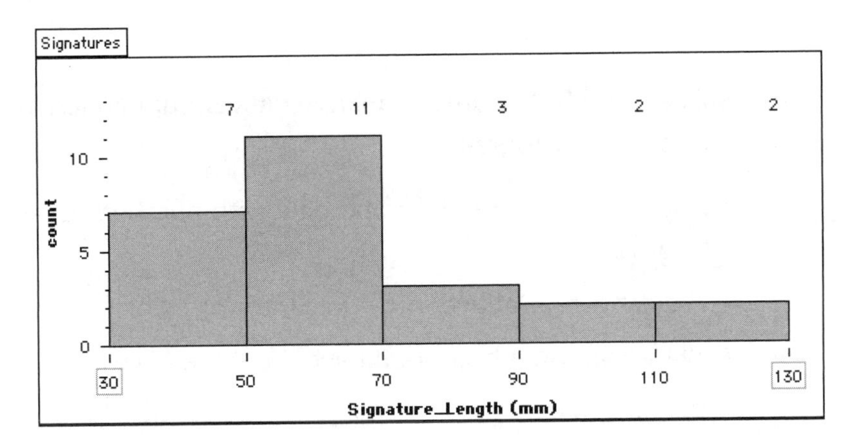

8. Answers will vary depending on bin width chosen. Students should notice that with smaller bin widths the distribution appears more jagged, with more ups and downs, whereas with larger bin widths it appears smoother. They should also notice that the distribution still has the same shape, with a large clump on the left side of the scale, tailing off to the right.

Histograms

Name:

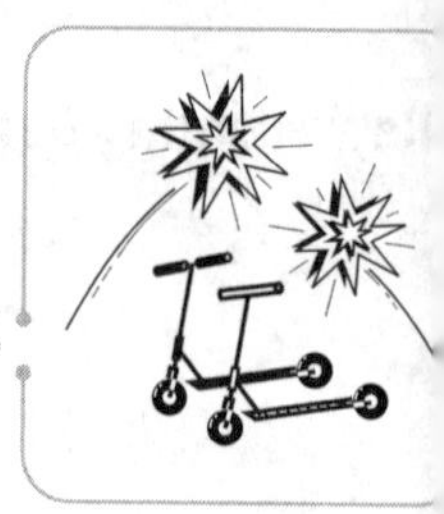

You will be using histograms to analyze data about injuries. These are some characteristics of histograms:

- Histograms are used to graph data with numbers as values, such as lengths.
- The data are grouped into equal intervals, which are called *bins*.
- All the bins on a histogram have the same *bin width*.
- The bins (bars) touch each other.
- An empty space means that there are no values in that bin.

HISTOGRAM OF AGES FOR SCOOTER INJURIES

Here is a histogram of the ages of 29 people who got injured riding scooters.

1. Where are the data for the injured 16-year-olds? Put an X on that bin.
2. Add data for these cases to the histogram. (You may need to add more bins to the histogram.)

   a. 5-year-old

   b. 24-year-old

   c. 33-year-old

   d. 35-year-old

   e. 44-year-old

   f. 50-year-old

COMPARE GRAPHS

Some doctors believe that students can hurt their backs by carrying heavy backpacks. Here are three different graphs of the same data—the weights of backpacks from 41 students in grades 5 and 7.

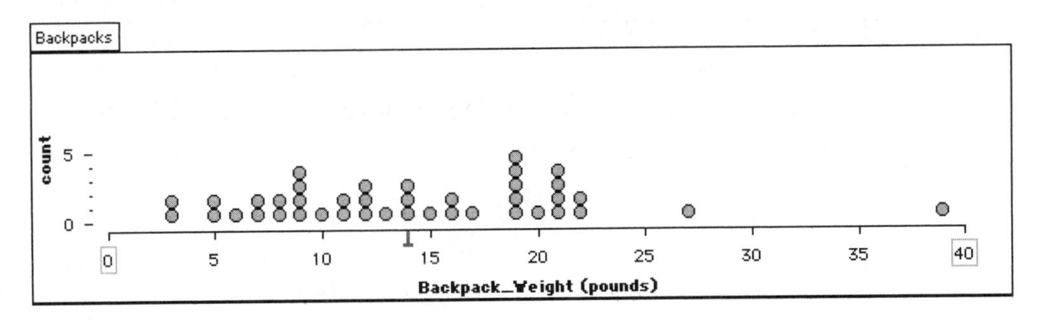

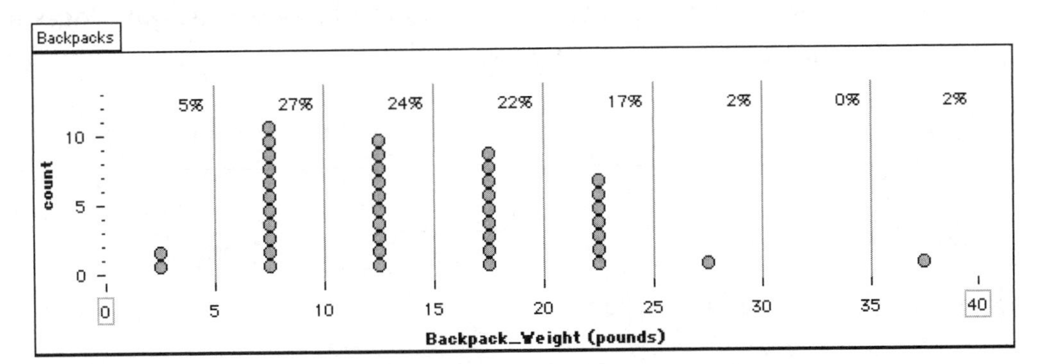

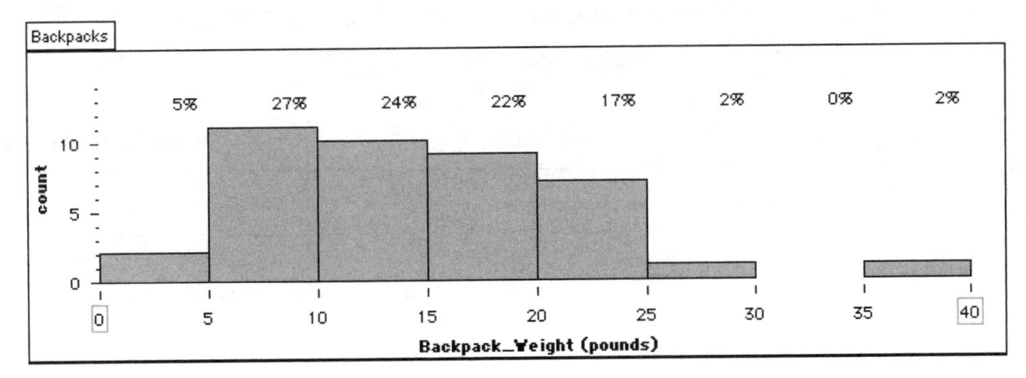

3. Where are the data for the backpacks that weigh 19 pounds? Put an X on each graph.

4. You have been asked to write an article about these backpack weights for the school newspaper. Answer these questions on a separate sheet of paper.

   a. List at least three points you would make to summarize the data.

   b. The newspaper has room for only one graph in the article. Which graph would you choose? Why?

Create Histograms by Hand

Name:

In this activity, you will create two histograms by hand and compare them.

FIND OUT ABOUT THE DATA

1. This data set has the measurements of 25 students' signatures in millimeters.

43 60 65 47 115 92 43 31 40 60 83 60 60

50 45 65 80 95 116 60 80 59 52 45 50

Put the signatures in order from shortest to longest length. Check to make sure you have 25 values.

GROUP THE DATA INTO BINS

2. For this histogram, use a bin width of 10. The first bin will be 30–39 mm. Figure out how many values fit in each bin. Fill out the table.

Bin	Values	Number of values
30–39	31	1
40–49	40, 43, 43, 45, 45, 47	6
50–59		
60–69		
70–79		
80–89		
90–99		
100–109		
110–119		

DRAW THE HISTOGRAM

3. Set up the scales for the histogram on graph paper.

   a. The scale for the horizontal axis is divided into *bins* of 10 mm. What scale will you use?

      30 to _________ mm

   b. The vertical axis has *counts* to show how many values are in each bin. What scale will you use?

      0 to _________

4. Make the histogram by drawing bars (bins) the way you would for a bar graph, except that the bins should touch each other. Leave an empty space only if there are no values in that bin.

CREATE A DIFFERENT HISTOGRAM OF THE SAME DATA

5. Now you will find out what happens when you change the bin width.

   a. Choose one bin width. Circle your choice.

      Bin width: 6 7 8 9 15 20

   b. How do you think the new histogram will compare to the one you made with bin widths of 10? Will there be more bins? Will the bins be taller?

6. Use a table to group the data from question 1, using your new bin width. Fill in the intervals for the bins in the first column. Then fill in the other two columns to figure out how many values are in each bin.

7. Graph the histogram on graph paper.

8. How do the two histograms you made compare?

Using Histograms and Circle Graphs to Analyze Data

OVERVIEW

The focus of this lesson is on using histograms to represent and analyze data about fireworks injuries. This is the first time in the unit that students create histograms with TinkerPlots. The software makes it easy to change bin widths, helping students see how different bin widths affect the way the data is represented. They need to select a bin width that provides a good overall representation of the data and that will enable them to answer questions for planning a public service announcement on fireworks safety. Students also use circle graphs to analyze data.

Objectives

- Analyze a large data set (100 cases)
- Create histograms to represent and analyze data
- Create, interpret, and compare circle graphs
- Use percentages to describe and analyze data
- Make recommendations that are based on data

TinkerPlots

Class Time: Two class periods

Materials

- Fireworks Safety, Part 1 worksheet (one per student)
- Fireworks Safety, Part 2 worksheet (one per student)
- Create Your Own Public Service Announcement worksheet (one per student, *optional*)

Data Set: Fireworks.tp (data for 100 people who got fireworks injuries)

TinkerPlots Prerequisites: Students should be familiar with intermediate graphing.

TinkerPlots Skills: Making histograms and circle graphs is explained in this lesson.

LESSON PLAN

Introduction

1. Introduce the fireworks injuries data set and its attributes. The data set has information about people who were injured by fireworks and went

to emergency rooms to be treated. As with the scooter injuries data set, this data was collected by the Consumer Product Safety Commission. Because the fireworks injuries data set has 100 cases, it provides a good opportunity for students to use histograms to represent and analyze a large data set.

2. Introduce the task of analyzing data to create a public service announcement (PSA) to prevent fireworks injuries. If you live in a state where consumer fireworks are illegal, ask students to create a PSA for states that allow fireworks. Discuss the goals of PSAs and give a few examples.

    - What are some examples of PSAs? [Some examples: "VERB: to get kids to be more active," "Got Milk?," and "Friends don't let friends drive drunk."]
    - Who you think is the target audience for each PSA? Is the PSA aimed at children, teens, adults, or people of all ages?

3. Go over the three main questions of the task. Make sure students are clear about the meaning of *target audience* for the PSA.

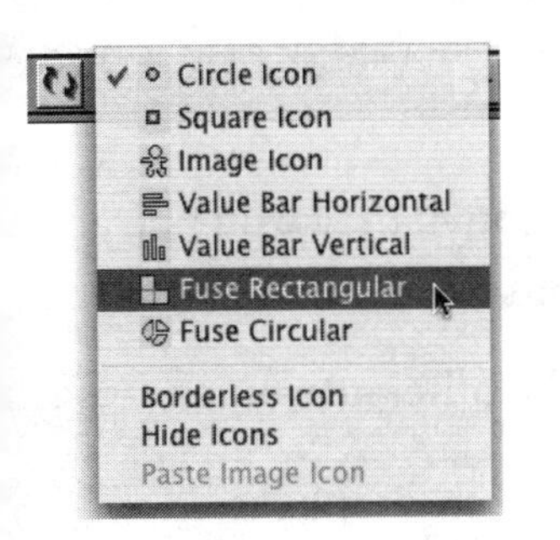

4. Because this is the first time students will make histograms in TinkerPlots, you may want to do a whole-class demonstration. To make a histogram, first separate the data by *Age* into at least two intervals. Then stack vertically. Finally, choose **Fuse Rectangular** from the **Icon Type** menu. To change the number of bins, you can drag a case right or left, as usual. You can also change the bin width by dragging the edge of a bin. Note that each case is still visible in the histogram. While you are demonstrating, you might go back and forth between circle icons and fused rectangular icons to help students see the relationship between interval plots and histograms.

5. Students need to decide which bin width best displays the data. Show students how to change the bin width by dragging the edge of a bin. Discuss how to choose a bin width for the histogram.

    - How does the shape of the distribution change as you increase the number of bins?
    - What do you see when the bins are very large? When the bins are small?

Exploration

6. Have students create their own histograms with TinkerPlots.

   Troubleshooting tips for histograms:

   - If you don't stack the icons, you will get floating rectangular fuses. To get a histogram, click the **Stack Vertical** button.

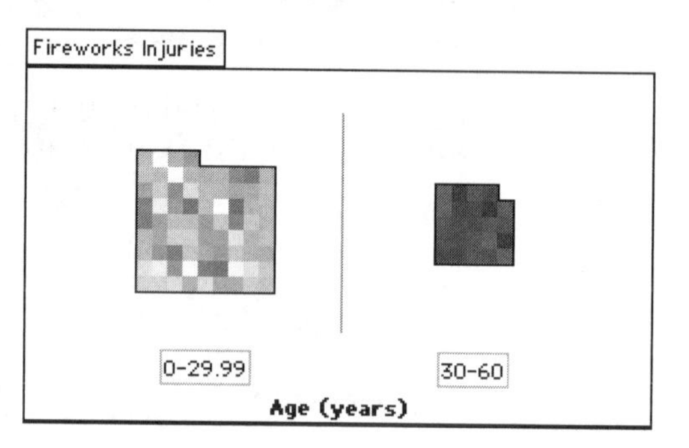

   - To order the cases within the bins, click the **Order Vertical** button.
   - You cannot create a histogram with only one bin by dragging the edge of a bin. (So when students start experimenting with bin widths, they need to drag to the left first.) You can make a graph with one bin by dragging a case in the right bin to the left, but note that *Age* no longer appears on the horizontal axis.

7. When students analyze their histogram in question 4, they won't be able to use dividers, but they can use percentages, and the median and mean.
8. After analyzing the age data, students analyze the other attributes.

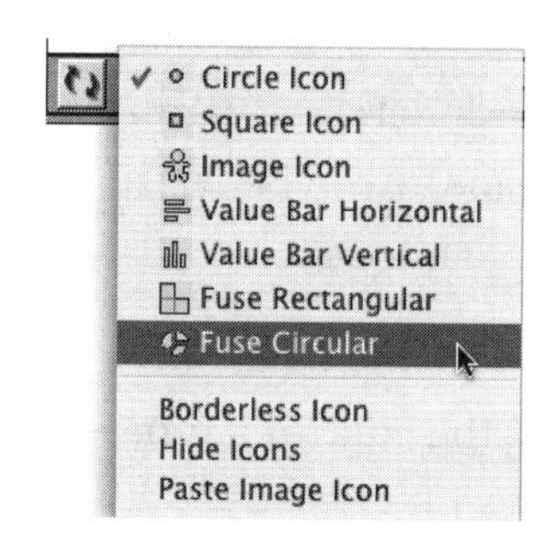

9. In the second part of the lesson, students make circle graphs to compare the frequency of the different types of injuries. You may want to demonstrate making circle graphs. Start with a mixed up plot (click the **Mix-up** button). Select the attribute in the data cards. Choose **Fuse Circular** from the **Icon Type** menu, then click either **Order** button. Click the **Key** button to see the categories.

   Troubleshooting tips for circle graphs:

   - You can make circle graphs starting from any other graph. If the circles are a solid color, select a different attribute in the data card to color them by something else.
   - You can separate circle graphs by categories as you would any other graph.

Wrap-Up

10. Discuss the two graph types students have been using.
    - How did you decide which bin width to use for your histograms? [This can be a difficult question for students. One helpful analogy is that if your bin width is too large it feels like you are standing so far away from the data that it's hard to see the distribution. And if your bin width is too small there are so many bars that it's hard to see the overall picture. It is like standing too close to the data.]
    - What are some strengths and weaknesses of circle graphs?
    - What features of data do you see better in a histogram? In a circle graph?
11. Have a class discussion about the findings.
    - What age group or groups should the public service announcement target? Why?
    - Should the PSA target males, females, or both? Why?
    - When in July should they show the public service announcement? Why?
    - What kinds of injuries are most common?
12. As an extension, you may want students to create their own PSAs intended to prevent fireworks injuries. Using the Create Your Own Public Service Announcement worksheet, they could create a short script for a PSA for TV or create something for a magazine. Have the class brainstorm criteria for high-quality PSAs and add criteria to the list in question 2.

    Some students may be confused about the goal of the PSA. It's helpful to clarify that the goal is to prevent fireworks *injuries,* not to prevent fireworks use. If consumer fireworks are banned in your state, then students could design the PSA for another state. A list of state regulations for fireworks is available at the Consumer Product Safety Commission website: www.cpsc.gov/cpscpub/pubs/july4/regs.html.

ANSWERS

Fireworks Safety, Part 1

4. A bin width between 5 and 10 is probably best for these data. Here are a few bin widths students might try.

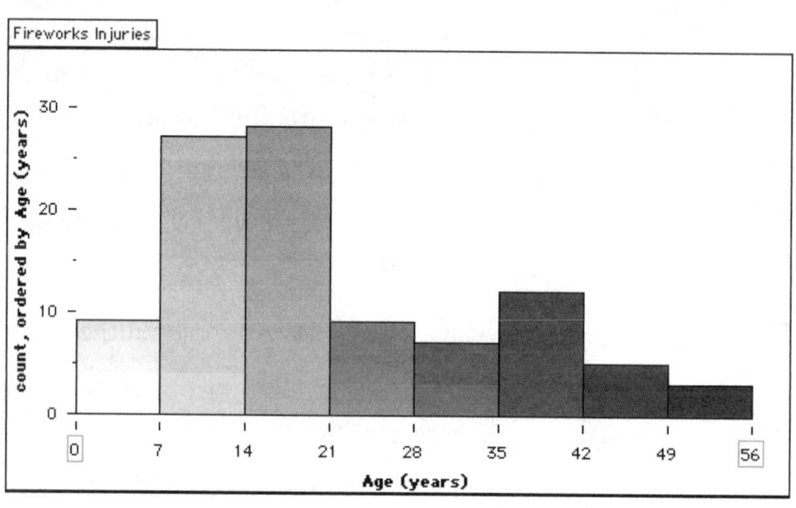

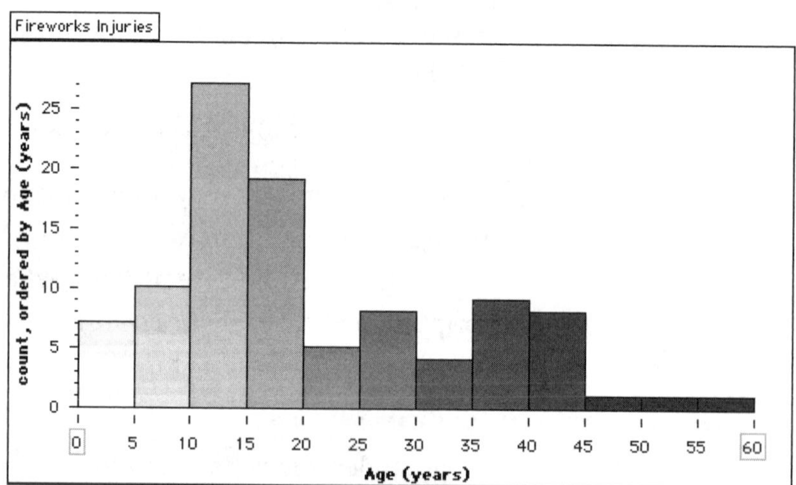

5., 6. Answers for question 5 will vary slightly depending on the bin width chosen. Sample student work:

Jessica: Target Age Group: 10–15. Because this is the age group that seems to have the most fireworks accidents. The other group that had a lot was the 35–45 year-olds. However, these are adults, and I think they probably had accidents. Gender: males. Because males tended to have the most injuries.

Comments: The student based her recommendations on the data, but could have used more specifics in her answer.

Emily: Target Age Group: 12–18 (teenagers). It seems that those are the ages where the most injuries are. It will also inform them not to use them in the future to prevent older injuries. Gender: males. Because young males usually seem to ignore warning labels more often then girls.

Comments: The student chose "males" based on personal experience rather than data.

Justin: Target Age Group: 10–15. The median was between 10–15. The peek [peak] was between 10–15. Gender: males. the majority of the people injured were boys. Around 75% were boys.

Comments: The student gave reasons for choices that were based on data. He used the terms *median* and *majority*, and gave percentages.

Briana: Target Age Group: 10–20 year olds. I think this is where most injuries occur, by seeing the huge clump in between the 10–20 range. Gender: both males and females. Even though lots of males get injured, some females do as well, so the warning should just go to everyone.

Comments: This student used the term *clump* in her reasons.

7. Answers will vary.
8. Sample answer: My first choice for when to show the PSA is July 2–5, and my second choice is July 3–6. The 4th and 5th are the most important days to cover: 38% of the injuries occurred on the 4th and 25% occurred on the 5th.

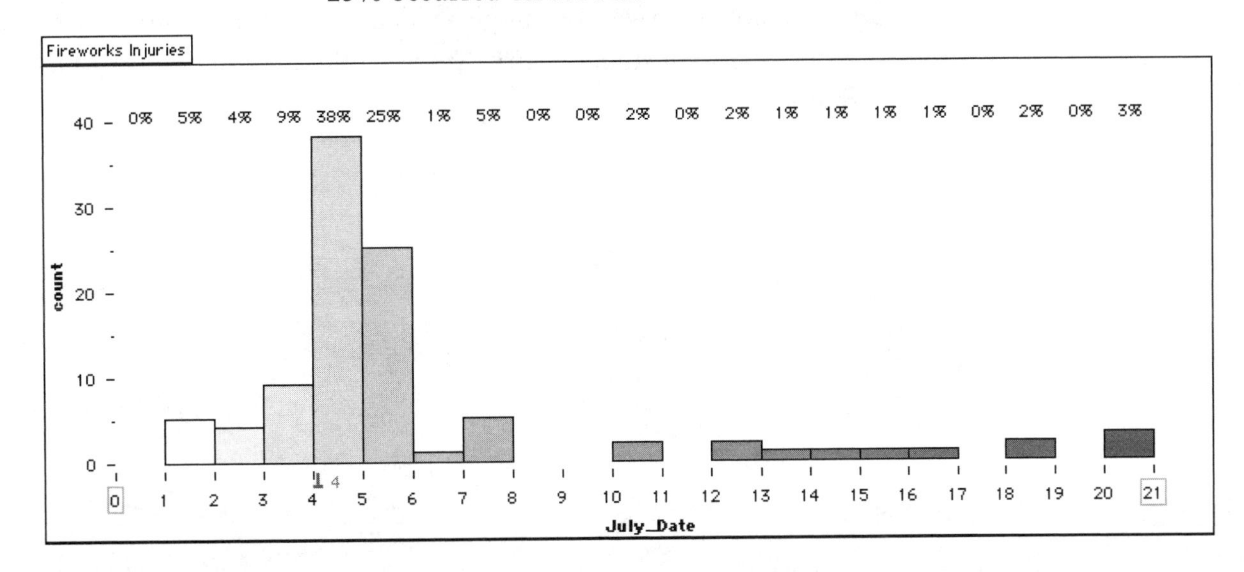

9. Yes, it does make sense to show the PSA after July 4th. Nearly half the injuries (44%) occurred after the 4th.

Fireworks Safety, Part 2

5. a. Estimates will vary.
   b. Eye: 52%; Face: 10%; Hand and Fingers: 28%; Leg: 4%; Trunk: 6%
   c. Sample answer: The most typical type of injury was to the eye (51% of cases). The next most common injury was to the hand and fingers (28% of cases).

7. Corrections will vary.
   a. No. More males (40) got eye injuries than females (12), but a higher percentage of females (60%) got eye injuries than males (50%).
   b. No. Over 8 times as many males (25) got hand and fingers injuries than females (3).
   c. No. A lower percentage of males (9%) got face injuries than of females (15%).

9. Corrections will vary.
   a. Yes
   b. No. 75% of the leg injuries were to males.
   c. No. More females (12) got eye injuries than face injuries (3).

Create Your Own Public Service Announcement

See the following pages for sample student PSAs. These PSAs were chosen to show a range of student work. Note that these are first drafts, and spelling and other errors have not been corrected.

THOMAS'S FIREWORKS PSA

FIREWORKS ARE COOL WHEN YOUR SAVE AND NOT A FOOL!

Injuries to Body Parts

40% eye, 28% arm + Hand, 14% face, 10% leg, 8% trunk

- Did you know that Most people Injer their eyes while playing with fireworks. Thats Why it is a good Idea to wear Sunglasses.
- Warning kids 10–15 Should be even More carefull since that is the age where Most kids get hurt.
- You Should always have an adult with you because they seem to get hurt alot less than kids
- Hey boys! lets try to get less injeries than girls since they beat us last time.
- Be exra carfull around July 4th because that is when Most people get hurt.

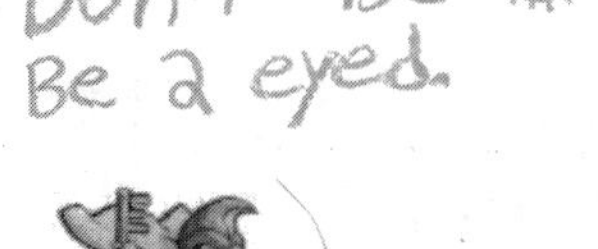

Don't Be ◎ eyed Be a eyed.

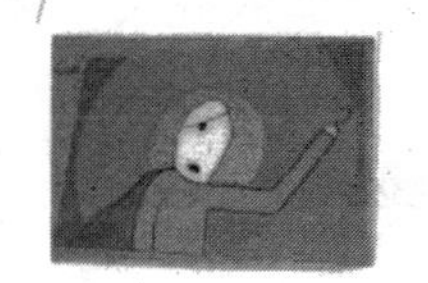

IAN'S FIREWORKS PSA

Be smart and safe with

Your doom does have to come with a boom
BE SAFE!

1. The most amount of injuries happen between the ages of 10 and 20
2. 52% of firework injuries are to the eye
3. Rockets are the most common kind of firework to cause injuries
4. There are much more injuries with males than females
5. The least dangerous firework that can cause injury is the foutain

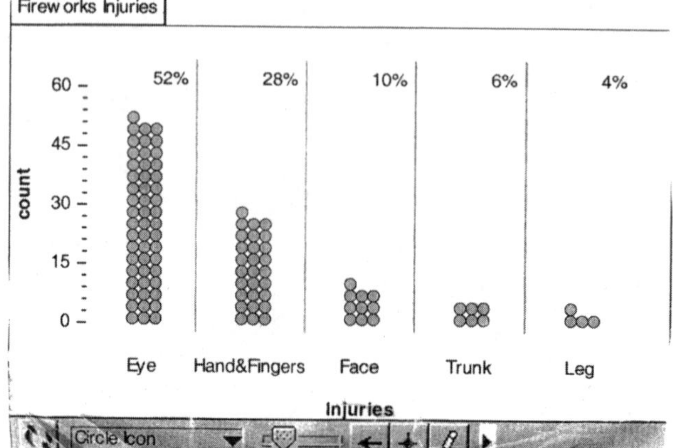

SU JUNG'S FIREWORKS PSA

Fireworks Safety, Part 1

Name:

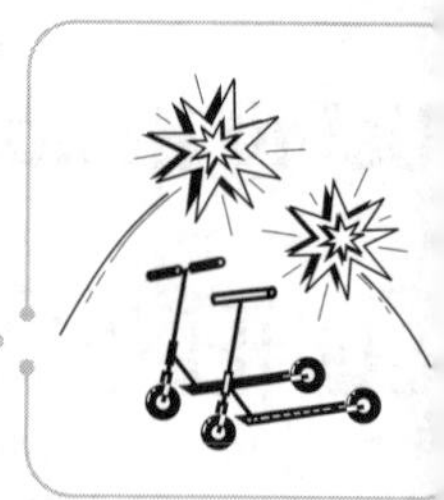

In this activity, you will use histograms and percentages to analyze data about 100 people who were injured by fireworks.

ASK A QUESTION

The Fireworks Safety Organization wants to launch a campaign to prevent fireworks injuries. They want to make a public service announcement (PSA) to be shown on TV.

The Fireworks Safety Organization needs to know who its target audience is. They want to make sure the announcement reaches the people who are most likely to be injured. (TV shows are designed to appeal to particular age groups—for example, children 5–10 years old). Your task is to analyze the data and make recommendations that answer these three questions:

- Who is typically injured by fireworks? (What age group? Males or females or both?)
- When in July should the PSA be shown?
- What kinds of injuries are the most common?

ANALYZE DATA AND COMMUNICATE CONCLUSIONS

Who Is Typically Injured by Fireworks?

1. Open the TinkerPlots file **Fireworks.tp.** Make different histograms of the age data by following the directions below.

2. Separate the age data into intervals and stack it. This example has two intervals.

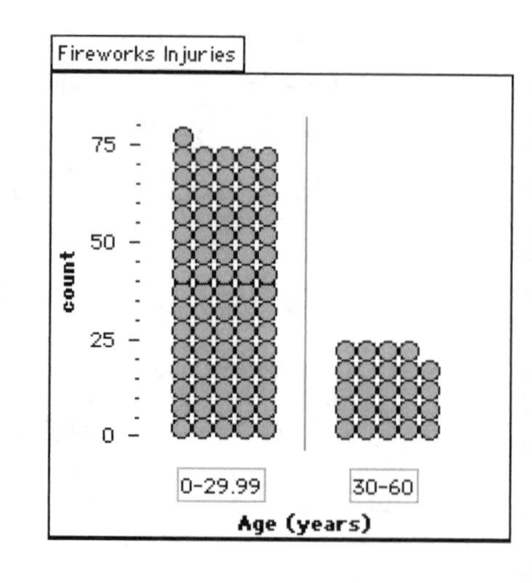

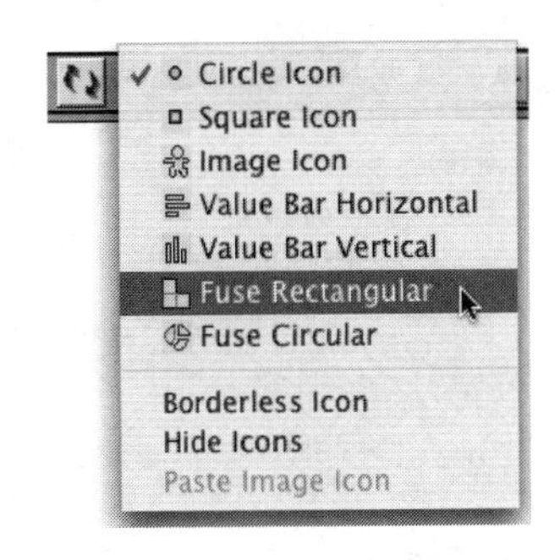

3. Choose **Fuse Rectangular** from the **Icon Type** menu on the lower plot toolbar. This shows the data as a histogram.

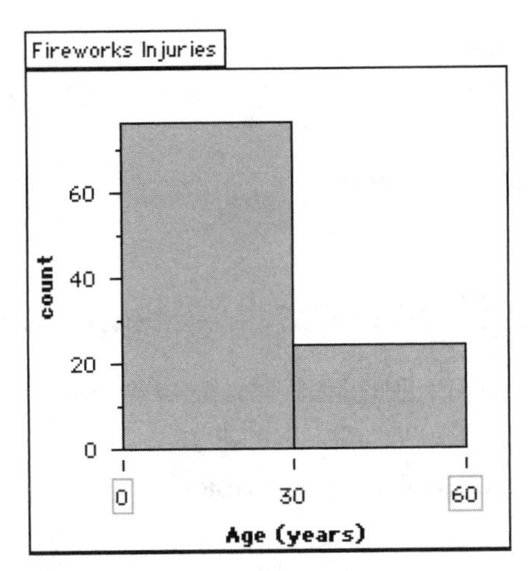

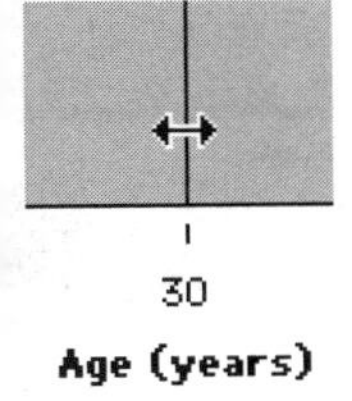

4. Experiment with creating different histograms by dragging the line between two bins. (Drag to the left to make more bins. Drag to the right to make fewer bins.)

   a. Try at least four different bin widths to find one that gives a good overall picture of the distribution of the data.

   b. What bin width do you think is the best for displaying the age data? ________

5. The organization wants to reach the people who are most likely to get injured. What age group or groups should the PSA target?

   Target age group or groups: ________

   Why? (Make sure to use the data to support your conclusions.)

6. Who should the PSA target in the age group that you picked? (Circle your choice.)

   Males   Females   Both males and females

   Why? (Make sure to use the data to support your choice.)

7. What TV shows should the PSA be shown on to reach the target audience? Why? (Answer this question based on your experience. This lesson does not have data on TV shows.)

When in July Should the PSA Be Shown?

8. The organization plans to show the PSA on TV for four days in a row. Based on the data, which four days do you recommend? Write your first choice and your second choice.

   July 1, 2, 3, 4 ________ July 2, 3, 4, 5 ________
   July 3, 4, 5, 6 ________ July 4, 5, 6, 7 ________
   July 5, 6, 7, 8 ________ July 6, 7, 8, 9 ________

   Why do you recommend these dates?

9. Do you think it makes sense to show the PSA at all after July 4th? Why or why not?

Fireworks Safety, Part 2

Name:

For the Public Service Announcement, you need to find out what kinds of injuries people typically get from fireworks.

CREATE CIRCLE GRAPHS

1. Open the TinkerPlots file **Fireworks.tp**.

   Follow these steps to make a circle graph of the types of injuries.

2. Select the attribute *Injuries* in the data card.

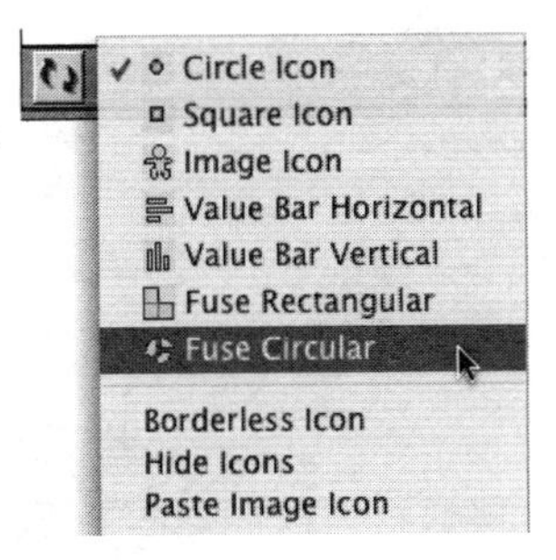

3. Click the **Mix-up** button on the lower plot toolbar, and then select **Fuse Circular** from the **Icon Type** menu. The dots will turn into one circle graph.

4. Click either **Order** button, then click the **Key** button to see what each color represents.

5. In TinkerPlots, the circle graphs do not show the percentages for each section.

   a. Estimate the percentage of injuries in each section of the circle graph.

      Eye: ______ Face: ______
      Hand and Fingers: ______
      Leg: ______ Trunk: ______

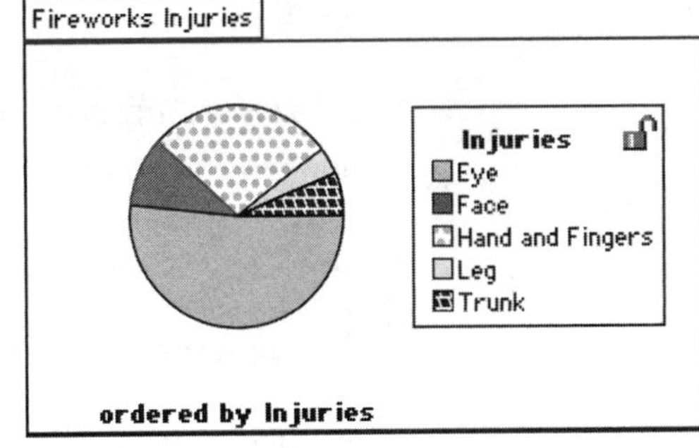

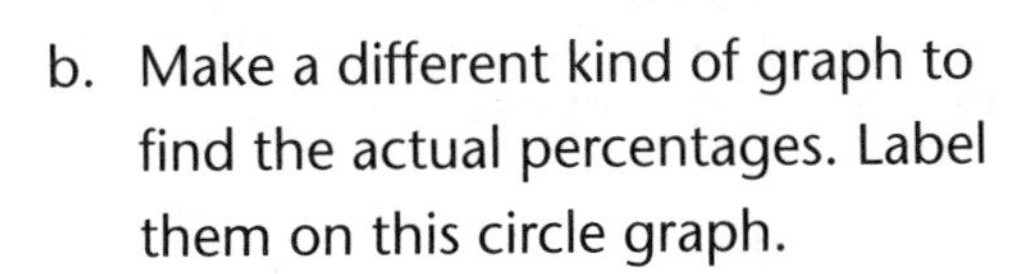

   b. Make a different kind of graph to find the actual percentages. Label them on this circle graph.

   c. For the PSA, write a summary statement that compares the frequency of different types of injuries. Include percentages in your statement.

COMPARE CIRCLE GRAPHS

How do the types of injuries that males and females got compare?

6. Make two circle graphs: one for males and one for females. Select the attribute *Gender* in the data card. Click either **Separate** button, then select the attribute *Injuries* again. This graph shows injuries for females.

7. Is each conclusion correct? Fix the incorrect ones to make them correct.

   a. More females got eye injuries than males.

   b. Twice as many males got hand and finger injuries than females.

   c. A higher percentage of males got face injuries than of females.

8. This circle graph shows what part of the 28 hand and finger injuries were to males and to females. Use Tinkerplots to make circle graphs like this one for all the types of injuries. (*Tip:* Drag *Injuries* to the horizontal axis, then select *Gender* in the data card.)

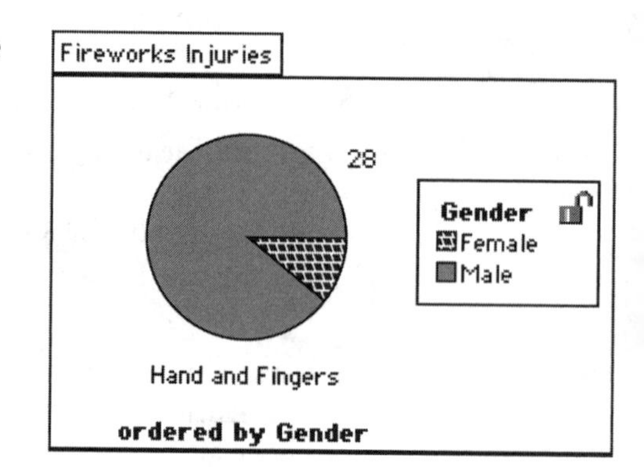

9. Is each statement correct? Fix the incorrect ones to make them correct.

   a. 25% of the leg injuries were to females.

   b. 75% of the males got leg injuries.

   c. More females got face injuries than eye injuries.

Create Your Own Public Service Announcement

Name:

Here is what the Fireworks Safety Organization wants: "We are looking for a public service announcement (PSA) to put in magazines or on TV in July. Come up with something that will appeal to the target audience. We want the PSA to tell people to use fireworks safely. We're not against fireworks—we just want to prevent injuries."

1. Decide what kind of public service announcement you will create. Circle your choice.

   A. Design a PSA for a magazine. You get a full page (8.5″ by 11″) to use for your PSA. Which magazine would you pick?

   ____

   B. Write a short script for a PSA that could be shown during a commercial break for a TV show. Which TV show would you pick?

   ____

   Note: If consumer fireworks are illegal in your state, design your PSA for use in another state.

2. Create your PSA. Here's what to include in the PSA:

   - At least five summary statements about fireworks injuries based on the data
   - A graph that provides information on fireworks injuries
   - *Optional:* A slogan for preventing fireworks injuries. An example of a slogan is "Got Milk?"

OVERVIEW

This lesson is designed to address a common difficulty that students have in writing about data. Some students make statements about what they see in the graphs, such as "there is a peak on the right side," without explaining what the statements means in the context of the data. In the activities, students improve example statements and write their own comparisons.

Objectives

- Interpret and compare histograms
- Write strong comparison statements
- Explain what statements mean in the context of the data

Offline

Class Time: One class period

Materials

- What Do the Statements Mean? worksheet (one per student)
- Two Histograms transparency (*optional*, on CD)

LESSON PLAN

Introduction

1. Introduce the task of comparing two histograms: the first on the ages of people injured by scooters and the second on people injured by fireworks. Show the Two Histograms transparency and ask the class questions to clarify how the graphs are set up.
   - What is the bin width in each histogram? [The bin width is 5 in both histograms.]
   - What do the percents at the top of each graph represent? [The percents represent the injured people in each bin.]
   - Why does it make sense to use percents instead of counts? [The number of cases in the scooter injuries is different from the number of cases in the fireworks data set. Comparing counts would be misleading. There are 28 cases of scooter injuries and 100 cases of fireworks injuries.]

Exploration

2. Have students work independently or in pairs to analyze the histograms. Then have them write three comparisons statements in response to question 1: How do the ages of people who got scooter injuries compare with the ages of people who got fireworks injuries?

3. Discuss the two histograms.

   - What similarities do you see between the ages of people who had scooter injuries and those who had fireworks injuries?
   - What differences do you see?
   - If the scooter injuries data set had 100 cases, how do you think the distribution of ages would compare to the fireworks injuries data set?

4. Introduce the last part of the activity, which is designed to help students improve their comparison statements. The table (for questions 2–7) has statements that were written by other students in response to question 1. These students made observations about the differences they saw between the two histograms. The observations are correct; however, the statements are weak because the students did not explain what their observations meant in the context of the data.

   Explain that the task is to write explanations for each of the statements to answer the question: What does each statement tell you about how the ages of the people compare for scooter injuries and fireworks injuries? Go over the example with the class.

5. Students work independently or in pairs to write explanations for the comparison statements.

Wrap-Up

6. Have students share the explanations that they wrote. One option is to have students work together with partners to improve the explanations. Ask students to consider these questions:

   - Is the explanation correct?
   - How could we make the explanation clearer and easier to understand?

7. Have a general discussion about writing strong comparison statements.
   - What suggestions would you give to these students to help them write strong comparison statements?

ANSWERS

What Do the Statements Mean?

1. Sample student work:

   Jessica: [1] Most people who get injured on a scooter or by fireworks are between 10–14. [2] There were many more people below 10 years old and above 15 years old who were injured by fireworks than by scooters. [3] The shape of the data for firework injuries was much more spread out than that of the scooter injuries.

   Comments: This student made a variety of comparisons, including comparing the shape of the data. The use of *most* is correct for scooter injuries but not for fireworks.

   Sean: [1] More for fireworks [2] Mod [mode] is bigger for scooter injuries [3] Range is lower for scooters.

   Comments: This student's work is a clear example of the type of weak statements this lesson is trying to improve. He did not explain what it means when the "range is lower" for scooters.

   Colleen: The most people who got hurt in both injury cases were between the ages of 10 and 15. There seem to be more firework injuries overall. While the data from the fireworks is spread out, the scooter injuries are not as much.

   Comments: In the first sentence, the student uses "most" incorrectly for fireworks. It's true that the mode is 10–15 for both graphs, but it is not the majority of people injured by fireworks. The student doesn't explain what "spread out" means in the context of the data.

2.–7. Sample answers:

What similarities or differences do you see in the histograms for age?	What does this mean about the ages of people who got fireworks injuries compared to the ages of the people who got scooter injuries?
2. Both graphs' highest peak is at the 10–15 interval.	**The highest peak means that people aged 10–15 yr were the most commonly injured by both fireworks and scooters.**
3. The bars for firework injuries are more spread out than the bars for scooters.	**This means that people from a wider range of ages got injured by fireworks than by scooters.**
4. The scooter injuries have a gap between ages 25 and 39. There are no gaps in the fireworks data.	**The gap means that no one between the ages of 25 and 40 yr got scooter injuries compared to 21% of the people who got fireworks injuries. Having no gaps for the fireworks injuries means that people in all the age intervals got injuries.**
5. There is an outlier in the scooter data but no outlier for fireworks.	**An outlier means that scooter injuries for people aged 40–45 yr are unusual. It stands out because there are no injuries for people ages 25–39 yr. In comparison, fireworks injuries for this age range are similar to the percent of injuries to people in the 25–40 yr age range (part of a trend in the data for ages 20–45).**
6. The data are more clumped on the left sides of both graphs than on the right sides.	**On the horizontal axis, the ages increase from left to right, going from 0 to 60 yr. More data on the left side means that injuries to younger people were more frequent than injuries to older people.**
7. The modes are the same for both.	**This means that the most common injuries for fireworks and scooters were to the same age group (10–15 yr).**

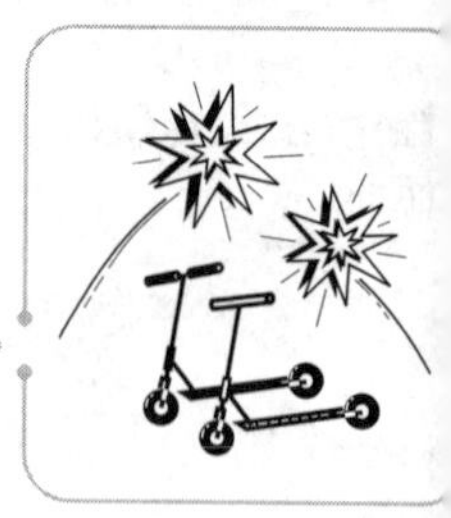

What Do the Statements Mean?

Name:

HOW DO THE AGES COMPARE?

1. How do the ages of the 28 people who got scooter injuries compare with the ages of the 100 people who got fireworks injuries? Write at least three comparison statements.

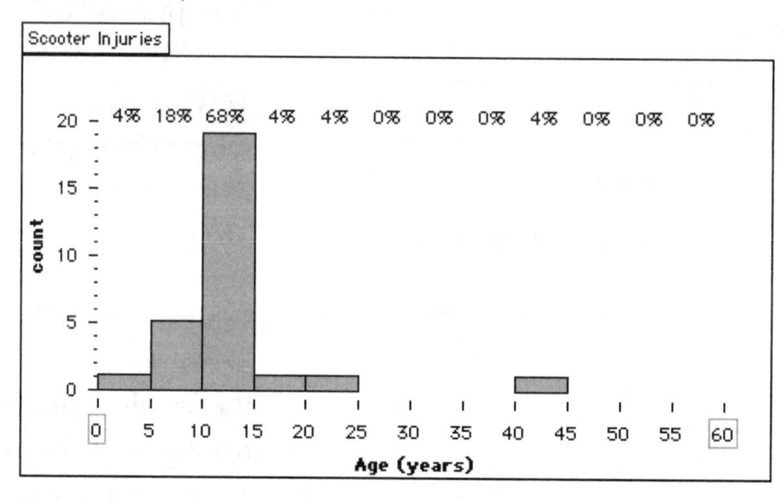

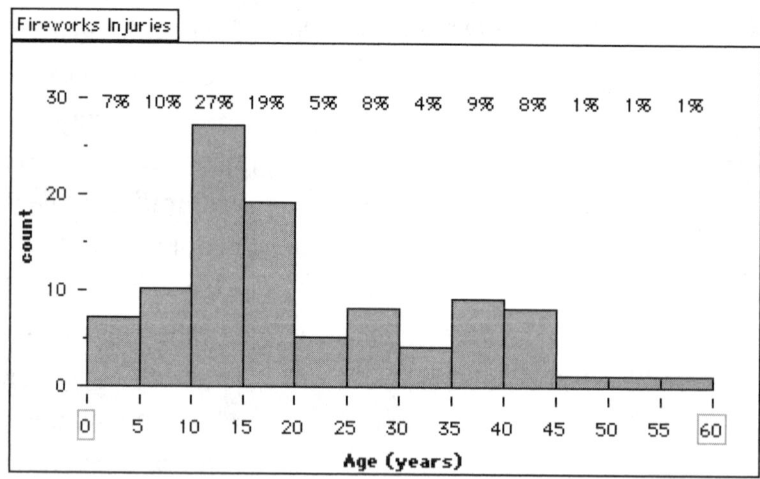

My comparison statements:

WHAT DID THESE STUDENTS MEAN?

Read each statement and write an explanation in the second column. Use the histograms to help you.

What similarities or differences do you see in the histograms for age?	What does this mean about the ages of people who got fireworks injuries compared to the ages of the people who got scooter injuries?
The fireworks have a bigger range.	A bigger range means that there is a larger difference between the highest and lowest ages of people who got fireworks injuries than the highest and lowest ages of people who got scooter injuries.
2. Both graphs' highest peak is at the 10–15 interval.	
3. The bars for firework injuries are more spread out than the bars for scooters.	
4. The scooter injuries have a gap between ages 25 and 39. There are no gaps in the fireworks data.	
5. There is an outlier in the scooter data but no outlier for fireworks.	
6. The data are more clumped on the left sides of both graphs than on the right sides.	
7. The modes are the same for both.	

8. On a separate sheet of paper, make a list of suggestions for writing strong comparison statements.

Section 5

Relationships between Attributes and Scatter Plots: Investigating Sports Data

In this section, students investigate the relationship between attributes by working with a new representation: scatter plots. To build students' understandings of this more complex representation, the lessons first focus on time-series graphs. Students learn how to read scatter plots from left to right and to look for overall trends. They learn how to draw informal lines of fit, and learn ways to analyze the type and strength of association between two attributes. Using these skills, students explore the relationship between attributes in data from basketball players in the Hall of Fame. In the final lesson, students investigate data from college basketball teams to determine whether they have a home-court advantage. They need to figure out what attributes to examine, how they will plot and analyze the data, and then prepare a presentation of their findings. The data in this section are drawn from track-and-field events at the summer Olympics and from basketball (both the Hall of Fame, and men's and women's college teams).

OBJECTIVES

Analyzing Data

- Create, interpret, and compare scatter plots (Lessons 5.1–5.5)
- Analyze changes over time (Lessons 5.1, 5.2)
- Identify overall trends in a group of cases (Lessons 5.1–5.5)
- Use informal methods to draw lines of fit (Lessons 5.3–5.5)
- Describe scatter plots by focusing on form, direction, and strength (Lessons 5.3, 5.4)
- Determine the type of association between two attributes (Lessons 5.3, 5.4)
- Determine what's typical for a group (Lessons 5.4–5.6)

- Create, interpret, and compare line plots (Lessons 5.4, 5.5)
- Determine which attributes to analyze in order to investigate a particular question (Lessons 5.5, 5.6)
- Use findings to pose new questions about data (Lesson 5.6)
- Use formulas to create new attributes (Lesson 5.6)
- Create and interpret a variety of plots to analyze one or more attributes (Lessons 5.5, 5.6)

Communicating about Data

- Write descriptions of scatter plots (Lesson 5.4)
- Write clear conclusions that use the data as evidence (Lessons 5.4–5.6)

Applications of Math Concepts from Other Strands

- Number and Operations: Apply and build understanding of ratios (Lesson 5.5)
- Number and Operations: Use percentages to analyze data (Lessons 5.4, 5.5)
- Number and Operations: Apply knowledge of positive and negative integers (Lesson 5.6)

TINKERPLOTS SKILLS

Students in this section use intermediate graphing skills: making and exploring graphs (such as grids, line plots, and scatter plots), and adding features to their graphs (such as dividers and percentages). They also create attributes using formulas, and learn to use a filter to look at a subset of cases. If your students have not used TinkerPlots before, you'll need to give them a basic introduction, using the movie *TinkerPlots Basics* in the TinkerPlots **Help** menu, and demonstrate some additional skills as needed, such as using dividers and formulas.

Working with Time-Series Graphs

OVERVIEW

Scatter plots are useful representations for analyzing the relationship between two quantitative attributes. However, scatter plots can pose several difficulties for students. Some students find it confusing that each point represents the values of two attributes. Unlike line plots, the points are not stacked on top of the axes, so it may seem like they are floating in space. This makes it more difficult to connect points with their values. Some students think that scatter plots look like a random splattering of dots and they are not sure how or where to examine them. In this lesson students learn how to read scatter plots by starting on the left side and moving across to the right side. Students begin by looking at the data one point at a time to see how Olympic winning distances have changed from year to year. Working with the attribute of years tends to make sense to students, perhaps because of their experiences with time lines in other subject areas. (This lesson is based on an exploration by Clifford Konold and is used with permission.)

Objectives

- Learn about the data set and the attributes
- Learn to read scatter plots by starting on the left side and moving across to the right side
- Interpret scatter plots to determine changes over time
- Analyze overall trends in data
- Consider the context of the data when interpreting the findings

TinkerPlots

Class Time: One class period

Materials

- How Have Olympic Results Changed over Time? worksheet (one per student)
- Name the Mystery Olympic Events worksheet (one per student)
- Meter stick or measuring tape (*optional*)

Data Sets: High and Long Jumps.tp (data for two Olympic events with a slider allowing students to see data one year at a time), **Olympic Gold Medals.tp** (data for 14 Olympic track-and-field events), **Mystery Events.tp**

TinkerPlots Prerequisites: None

TinkerPlots Skills: Using sliders is explained in this lesson.

LESSON PLAN

Introduction

1. Introduce the context of looking at data from the Olympics (1896–2004) for three men's events: the high jump, long jump, and 200-meter race. Go over how these events work so that students are familiar with them before analyzing the data. The results for the high jump and long jump are in meters. To give students a sense of the heights, hold up a meter stick and ask students to estimate the height of the classroom door or wall in meters. The high jump for 1896 was 1.81 meters, or about 6 feet. Students could also measure the length of the floor in meters to get a sense of the long jump distances.

2. Introduce the first question: How have the gold medal winning heights for the high jump changed over the years? Ask students to write down their hypotheses. Students are likely to say that the heights have increased over the years. You may want to ask some questions to help students get more specific about their hypothesis. For example: Do you think the heights increase each Olympics from 1896 to 2004? Do you think the heights increase consistently by a little each Olympics or by large amounts for some Olympics?

3. Open the file **High and Long Jumps.tp.** Demonstrate how to use the slider to show the high jump data for one year at a time. Explain that the position of each point is determined by two values: the year of the Olympics and the height of the winning high jump. Ask students to describe how the winning heights are changing from year to year.

Drag the pointer slowly across the years. Dots will appear in the scatter plot. You can also animate the slider by clicking the arrow, but the animation moves too fast for this lesson. Students should use the pointer.

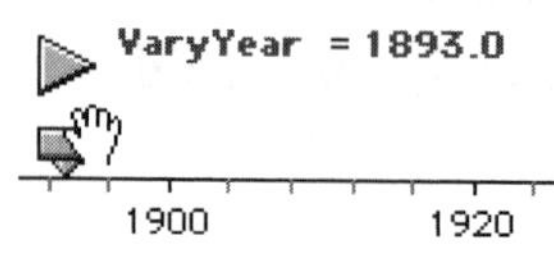

Exploration

4. Have students work with partners to do the high jump activity (questions 1–4). One student moves the slider a year at a time, and the other talks about how the data is changing. The talking student has the role of a "sports announcer" who is describing the results for this event over time. This activity helps students learn to read scatter plots by starting on the left side and looking across the points to the right side.

Troubleshooting tips for sliders:

- Students need to start with the pointer at the earliest year.
- Students should use the pointer so they can move slowly from year to year and have enough time to talk about how the results are changing. Because the animation moves too fast, they should not use the arrow.

5. After looking at the changes from year to year, students need to shift to looking at the overall trend of the data (question 5). While the high jump heights went down in some years, the general trend is that the heights have been increasing.
6. Students then do a similar activity for the data on long jumps (questions 6–8). They first use the slider to look at and describe the changes from year to year. Then, they analyze the scatter plot to determine the overall trend of the data.
7. Have a short class discussion about the data for high and long jumps.
   - What is the overall trend of the data for the high jumps? How can you tell from the scatter plot?
   - How much have the heights changed over time?
   - What is the overall trend of the data for the long jumps? How can you tell from the scatter plot?
   - How much have the distances changed over time?
8. The next part of the activity involves investigating a larger data set with more Olympic events. The central question remains: How have the gold medal results changed over time? Students use TinkerPlots to create a scatter plot of the winning times for the men's 200-meter race. As you circulate around the room, check to see whether students understand that the times are decreasing means the runners are faster. Some students assume that if values decrease, then they are getting worse, without considering the context of the data.

   Troubleshooting tips for scatter plots:

   - Students need to fully separate the data in two directions—horizontally and vertically. It's helpful to first make a line plot of *Year* on the horizontal axis and then to put the other attribute on the vertical axis and fully separate the data vertically.
   - Students can use two reference lines (one horizontal and one vertical) to help them read the values for points.

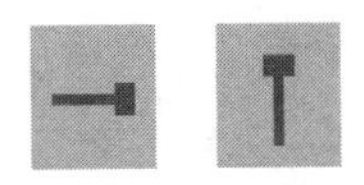

9. In the Name the Mystery Olympic Events activity students need to figure out which Olympic event is represented by each scatter plot. (This is similar to the mystery attributes activities in Lessons 1.2 and 2.2.) Have students open the file **Mystery Events.tp.** They'll need to scroll down to see the different graphs. They can also move the data card by selecting it and dragging the top border. The main goal is for students to interpret each scatter plot in the context of the Olympics and consider whether it is more likely to represent winning distances or winning times. The activity also gives students practice in making scatter plots with TinkerPlots. Here are some questions you can ask if students have difficulty figuring out the mystery events.
   - Do you think the scatter plot represents winning times or winning distances? Why?
   - What clues can you get from the direction of the trend?
   - What clues can you get from the values on the vertical axis?

Wrap-Up

10. Go over the answers to the Name the Mystery Olympic Events activity. Have a class discussion about students' strategies and the general trends.
    - What strategies did you use to figure out which Olympic event each scatter plot represents?
    - For each event, what is the overall trend of the data?
    - Which event has had the most consistent improvement over time? How can you tell from the scatter plot? [Men's pole vault has had the most consistent improvement—there are no points very far away from the others.]
    - Which event has had the greatest improvement over time? Why? [For the women's javelin throw, the distance has nearly doubled.]

ANSWERS

How Have Olympic Results Changed over Time?

1. Answers will vary. Students will probably think the gold medal heights have increased.

5. Sample answer: The upward trend was irregular at first, then, except for one year, the gold medal height increased from 1948 to 1976. Since 1980, heights have stayed within 0.15 m.

8. Sample answer: Since 1900 the long jump distances have mostly been gradually increasing. There are drops in 1920, 1948, and 1952, and a sharp increase in 1968.

12. Sample answer: Times varied during the early 1900s. They decreased from 1932 to 1978 except for unusually short times in 1936 and 1968. Since 1988 times have gone up and down.

Name the Mystery Olympic Events

2. Men's pole vault
3. Women's 100 meters
4. Men's 10K
5. Women's javelin
6. Sample answer: First I thought about events with either a trend up or a trend down. If there were several years without data, I tried women's events that would match the direction of the trend.
7. Sample answers:

   Olympic Event A: Until 1988 the distance was equal or better than the previous year, with only two exceptions, 1920 and 1948. Since 1988 the trend has leveled off at about 6 m with equal ups and downs.

   Olympic Event B: The first three years that women's 100 meters was an event, the times decreased dramatically. The time increased in 1948. Since then, the times have decreased or fluctuated around 11 s.

   Olympic Event C: The time for the men's 10K went up in 1920 and 1968, but the times in general have been decreasing. The time now fluctuates around 27 min.

   Olympic Event D: The distances that women threw the javelin increased until 1988, except for a small drop in 1968. The 1988 distance has not been beaten as of 2004.

How Have Olympic Results Changed over Time?

Name:

You will learn how to interpret scatter plots that show changes over time.

MEN'S HIGH JUMP

In the high jump, the competitor who jumps over a bar at the greatest height wins. You will investigate the question: How have the Olympic gold medal heights for the men's high jump changed over time?

1. In 1896, the gold medal height for the high jump was 1.81 meters. Make a hypothesis: How do you think the gold medal heights have changed over the years from 1896 to 2004?

2. Open the TinkerPlots file **High and Long Jumps.tp** to see the gold medal heights for the men's high jump events in the Olympics.

3. Use the slider and the first graph to show the high jump results for one Olympics at a time. To use the slider, drag the pointer slowly from year to year. The dots will appear on the scatter plot.

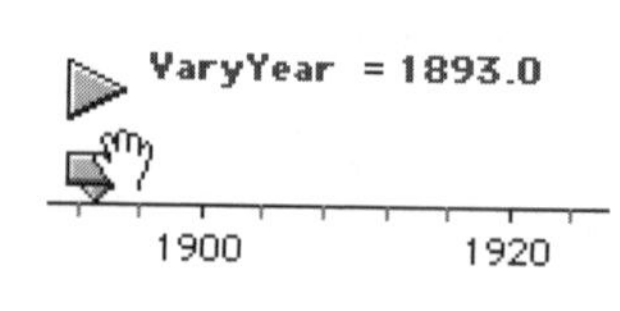

4. Work with a partner: One person moves the slider starting at 1896, one year at a time, and the other person talks about how the heights change from one Olympics to the next. You might say "The height went up a little," or "It went down again," or "It stayed the same."

5. Now, look for overall trends in the data. Move the slider to 2004 to show all the points on the scatter plot. Interpret the scatter plot by starting on the left side and moving across to the right side. How did the heights change over time? What is the overall trend of the data?

MEN'S LONG JUMP

Now, you will analyze the data for the attribute *Long_Jump*. The competitor who jumps the greatest distance is the winner.

6. Scroll down to the second graph.
7. How did the gold medal distances for long jumps change from one Olympics to the next? Use the slider to display one dot at a time. Tell your partner how the distances changed.
8. Move the pointer to 2004 to display a scatter plot with all the points showing. How did the long jump distances change over time? What is the overall trend of the data for long jumps?

MEN'S 200 METERS

Next you will analyze the gold medal times for the men's 200-meter race. The competitor who runs the race in the shortest time is the winner.

9. Close the High and Long Jumps data set and open the file **Olympic Gold Medals.tp** to see the gold medal times and distances for different events.
10. Put *Year* on the horizontal axis and fully separate the data to make a line plot.
11. Put *M_200_Meters* on the vertical axis and fully separate the data vertically to make a scatter plot.
12. Analyze the data: How have the gold medal times for the men's 200-meter race changed over the years?

Name the Mystery Olympic Events

Name:

For questions 2–5, figure out which Olympic event is represented by each scatter plot. Then, make the graph with TinkerPlots.

Tip: Scroll down to see the different graphs. You can also move the data card by selecting it and dragging the top border.

1. Open the TinkerPlots file **Mystery Events.tp.**
2. What is Olympic Event A? ____________________
   Make the scatter plot with TinkerPlots.
3. What is Olympic Event B? ____________________
   Make the scatter plot with TinkerPlots.
4. What is Olympic Event C? ____________________
   Make the scatter plot.
5. What is Olympic Event D? ____________________
   Make the scatter plot.
6. How did you figure out which Olympic events the scatter plots represented?
7. Choose a scatter plot for one of the Olympic events. Circle your choice: A B C D. Analyze the data. How have the results for this event changed over time?

Making and Analyzing Scatter Plots

OVERVIEW

In this lesson, students are given a table of data and need to make a scatter plot by hand. They select scales for the axes and then plot the points. Creating the scatter plot helps students understand that each point represents the values of two attributes. Building on the previous lesson, students analyze scatter plots to determine how the winning Olympic times or distances have changed over time. They also make predictions for future Olympics and sketch scatter plots to match descriptions.

Objectives

- Set up the scales for a scatter plot and plot the points
- Interpret scatter plots to determine changes over time
- Analyze overall trends in data
- Create scatter plots to match descriptions
- Apply knowledge of decimals to plotting points on a scatter plot

Note: The lesson assumes that students have had some experience setting up scales for graphs.

Offline

Class Time: One to two class periods

Materials

- Women's 200 Meters worksheet (one per student)
- Men's Discus worksheet (one per student)
- Grid for Making Scatter Plots transparency (on CD)
- Winning Times for Women's 200 Meters transparency (on CD)
- Winning Distances for Men's Discus transparency (on CD)
- Graph paper
- Graph paper transparencies (*optional*)
- Rulers or index cards (*optional*)

LESSON PLAN

Introduction

1. Introduce the goal of creating and interpreting scatter plots by working with more Olympic data. Students will continue to explore the question from the previous lesson: How have the results for different Olympic events changed over the years?

2. The first part of the activity involves making a scatter plot of data from the women's 200-meter race. If students already know how to make scatter plots, they can work independently on the task in class or for homework.

   If students have little or no experience with making scatter plots, it's helpful to go through the steps with the class. Display the Grid for Making Scatter Plots transparency and demonstrate how to set up the scales and plot the points.

   Setting up the scales:

   Label the horizontal axis with *Year* and explain that it is the convention for time-series graphs to put the years on this axis. Then label the vertical axis with *Winning Time (seconds).* Use the questions to help the class figure out the two scales.

   - What scale would you use for *Year*? Why? [A scale of 1948–2004 in 15 four-year intervals works well because the summer Olympics take place every four years.]
   - What are the units for the winning times? [seconds] Why are they measured to the tenths place? [The race is very fast, and runners' times can be very close together.]
   - What are the minimum and maximum values for the winning times? [21.3 s, 23.0 s]
   - What scale could we use for the winning times? [Some possibilities: 21.0–24.5 in increments of 0.5 or 21.0–24.4 in increments of 0.2.]

   Show students how to put hatch marks on the axes to show that they don't meet at 0.

Plotting points:

Demonstrate how to plot the first few points on the scatter plot. Show how each point represents two values: one for year and one for winning time. Then, have students plot the points on their own scatter plots. They can use rulers or index cards to help them figure out the positions of each point.

Checking the scatter plot:

Ask students for suggestions of how to check the scatter plot. Here are some questions that students could use to check their work.

- Does the scale increase in equal intervals? Are any numbers missing from the scale?
- Does the number of points on the scatter plot match the number in the table?
- Do the values of the points match the values in the table?

3. After students have completed their scatter plots, discuss these questions.
   - How have the winning times changed over the years? What is the overall trend of the data? [They went down until 1988, but now are staying near 22 s.]
   - Are there any exceptions to this trend? [1988 was unusually low.]
   - What are some things to remember about making scatter plots?
   - What are some mistakes that people might make when they create a scatter plot? What are ways to avoid those mistakes?

Exploration

4. Before students work with the discus data, make sure they understand the event. The winner is the competitor who throws the discus, a heavy disc, the longest distance. The discus for the men's event weighs 2 kg, or about 4.5 lb. Have students work independently on questions 1–3 on the Men's Discus worksheet.
5. Have a class discussion about the answers to questions 1–3. It's helpful to display a transparency of the scatter plot so that students can refer to it during the discussion.

- Which year had the shortest distance? The longest distance?
- How have the winning distances changed over time? What is the overall trend of the data?
- What are your predictions about the winning distances in 1940 and 1944? Why?
- What are your predictions for the winning distances for the next four Olympics? Why?

6. Introduce question 4, which involves creating scatter plots to match descriptions. You may want to go through an example with the class before having students work independently.

Wrap-Up

7. Have a class discussion about the scatter plots that students created. You may want to have a few students draw their scatter plots on a grid transparency.
   - How did you figure out how to position the points to match each description?
   - What did you write for your own description? How did you draw a scatter plot to match it?
8. Discuss how other attributes have changed over time.
   - Which attribute did you pick? How do you think it has changed over time? Why?
   - What would a scatter plot of your attribute and the attribute *Year* look like? Why?
   - What are some other examples of attributes that have values that have increased over the years? Decreased over the years?

ANSWERS

Women's 200 Meters

1.–4. Sample scatter plot:

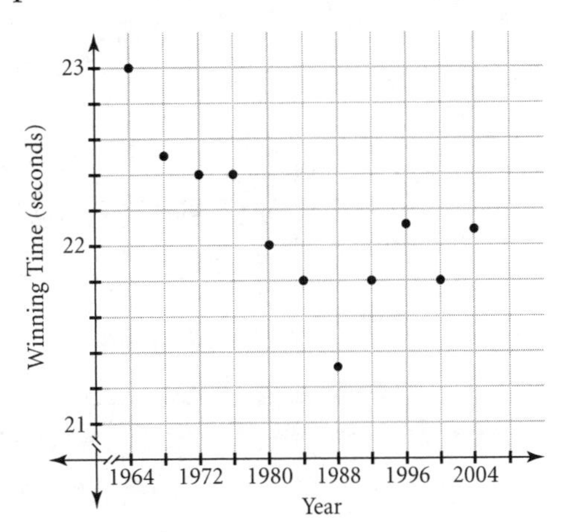

5. Times have gone down since 1948, but since 1992 they have leveled off at about 22 s.

Men's Discus

1. a. 1896; 29 m

   b. 2004; about 70 m

   c. There is a sharp increase in distance from 1896 to 1912, then a slower increase until 1968, then a slight, inconsistent increase to 2004. Overall, the distances have consistently increased.

2. Sample answers:

   a. 1940: 51.3 m; 1944: 52.0 m

   b. Distances were increasing at about 0.7 or 0.8 meters per year.

3. Sample answers:

   a. 70.1, 70.2, 70.3, 70.4 m

   b. The upward trend has slowed over the past three Olympics and will probably get slower.

4. Sample answers:

a.

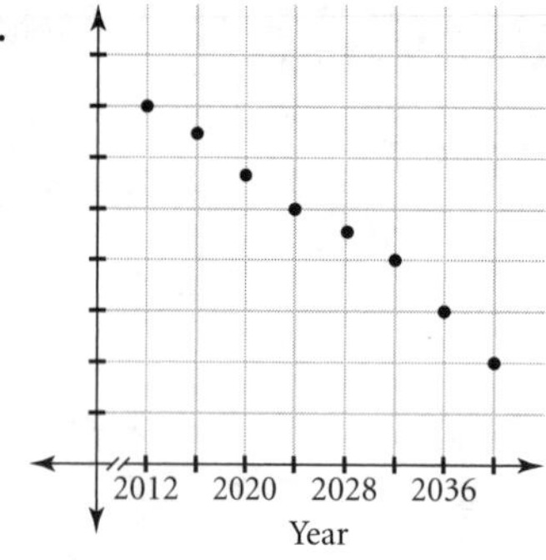

b.

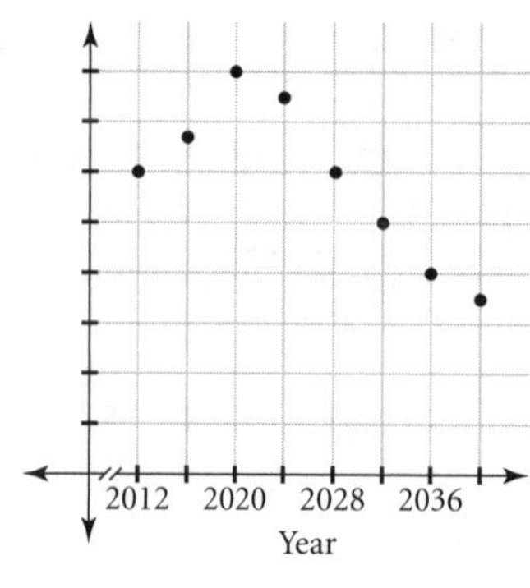

c.

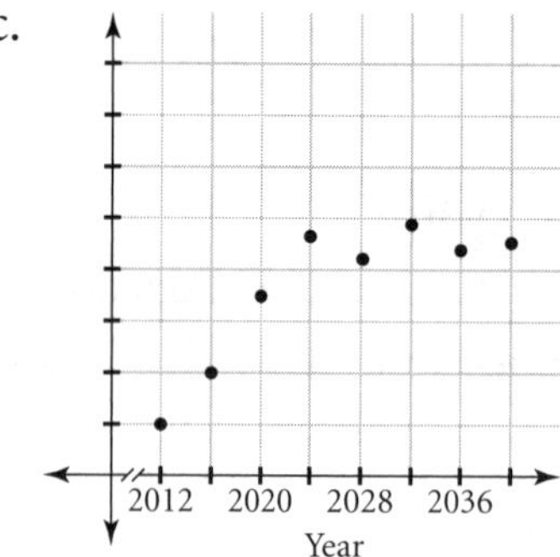

d. The gold medal distances stayed about the same for awhile, then they increased a lot.

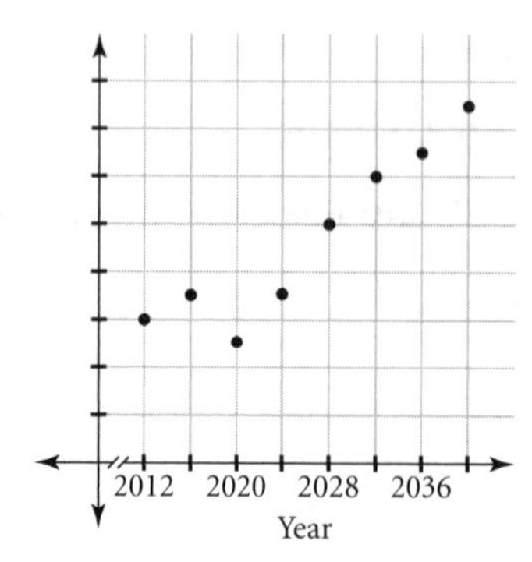

5. Sample answers:

   A. The average height of U.S. adult males has increased over the last 200 years, more rapidly during the first 100 years. It is not increasing very much now.

   B. The cost of a personal computer has gone down since the early 1980s. The decrease in price was fastest during the first 10 to 15 years computers were available.

   C. During the first 100 or more years of U.S. history, a big family was an advantage for help with a farm. Since then the size has decreased.

   D. The number of people using the Internet was small during the 1980s, when it was used only by governments and universities. As computer prices went down and Internet access became more available and cheaper, the number of people using the Internet increased rapidly.

Women's 200 Meters

Name:

This table shows the gold medal winning times for the women's Olympic 200-meter race (rounded to tenths of a second). Follow the steps below to make a scatter plot of the data.

Year	Winning time(s)
1964	23.0
1968	22.5
1972	22.4
1976	22.4
1980	22.0
1984	21.8
1988	21.3
1992	21.8
1996	22.1
2000	21.8
2004	22.1

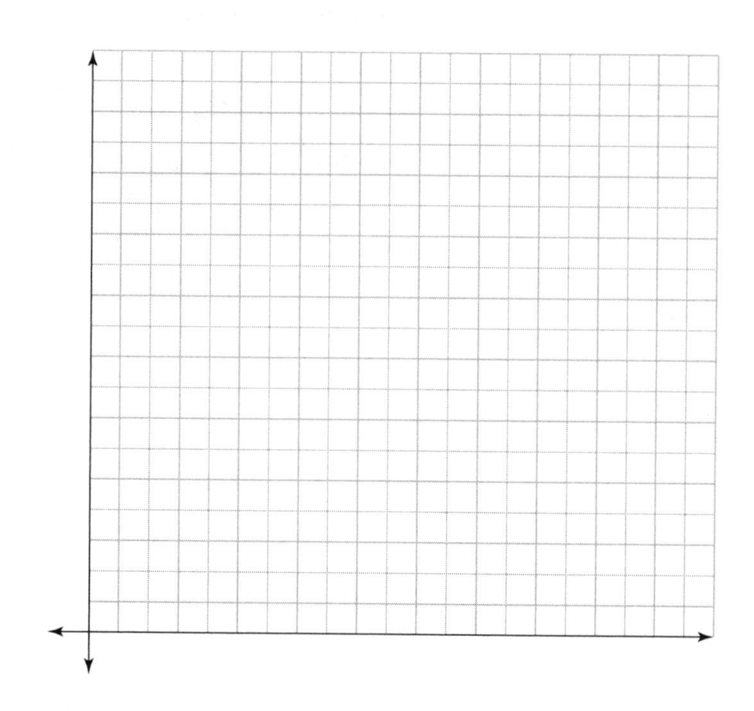

1. Label the horizontal axis with *Year.* Label the vertical axis with *Winning Time* (*seconds*).

2. Use the questions below to help you choose your scales. Then put your scales on the two axes.

   - What are the minimum and maximum values? What is the range?
   - What intervals will you use on the scale? Remember that the values on a scale need to increase in equal intervals, such as 1.2, 1.4, 1.6.

3. Plot the points on the graph. Remember that each point corresponds to two values: the year and the winning time in seconds.

4. Add a title to your scatter plot and check the scales, the number of points plotted, and the positions of the points.

5. How have the winning times changed over the years?

Men's Discus

Name:

1. The scatter plot below shows the gold medal distances for the men's discus event.

   a. When was the distance the shortest?
   Year: ________ Distance: ________ m

   b. When was the distance the longest?
   Year: ________ Distance: ________ m

   c. How have the distances changed over time? What is the overall trend of the data?

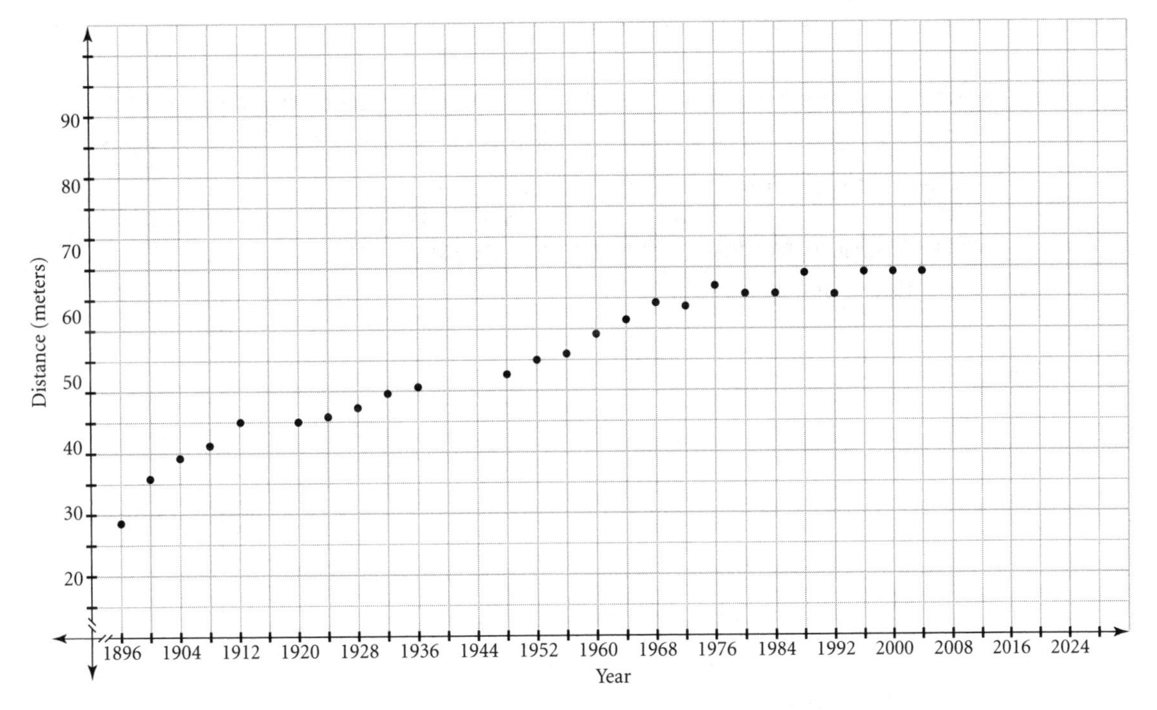

2. There were no Olympic games in 1940 and 1944. The Olympics were not held because of World War II.

   a. Based on the trend of the data, what do you think the gold medal distances would have been for those years?

   1940: ________ m 1944: ________ m

   Add your points to the scatter plot.

b. What are your reasons for making these predictions?

3. Think about what the results might be for future Olympics.

   a. Based on the trend of the data, what do you think the gold medal distances will be for the next four Olympics? Add those points to the scatter plot.

   b. What are your reasons for making these predictions?

4. These scatter plots have a scale of years for eight Olympic games from 2012 to 2040. On each scatter plot, draw eight points (one for each Olympic game) to match the descriptions.

   a. The gold medal times improved steadily over the years.

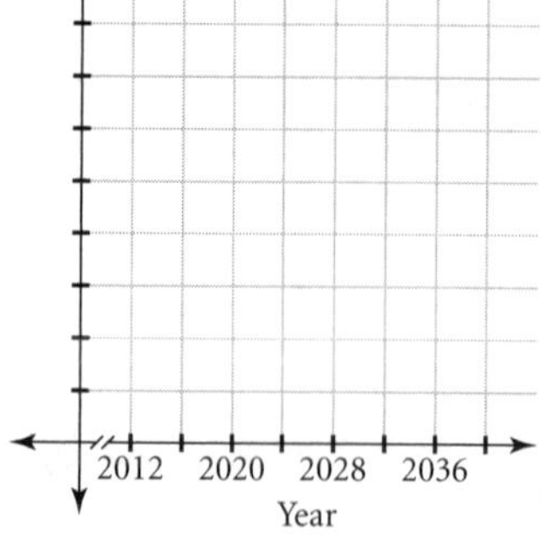

   b. The gold medal times got worse for several Olympics, but then they improved.

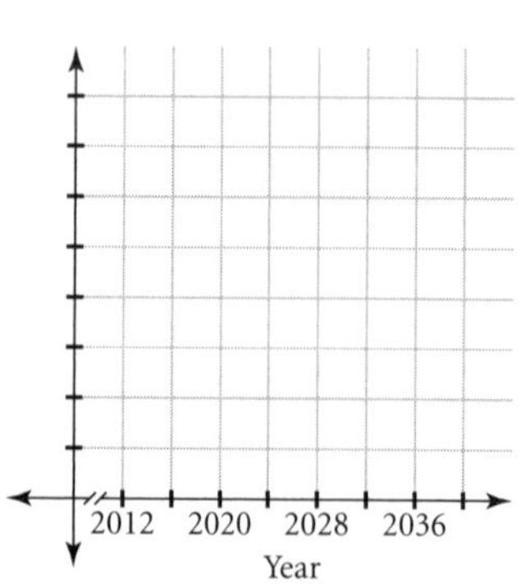

c. The gold medal distances improved for several Olympics, but then they stayed about the same.

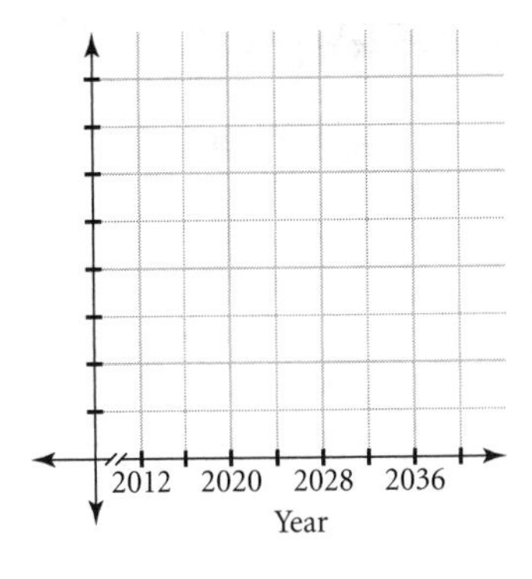

d. Write your own description, then draw a scatter plot to match it.

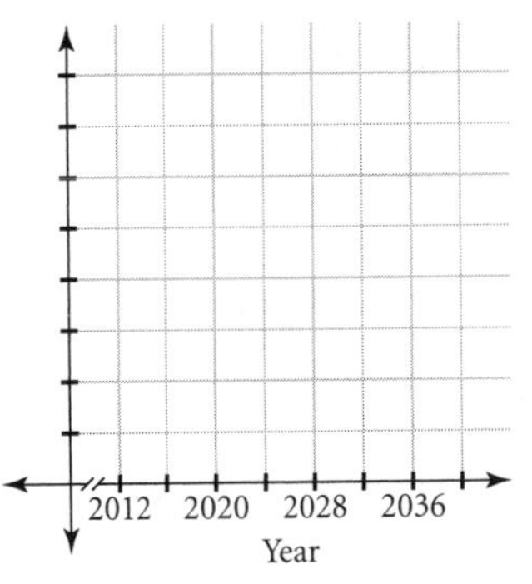

5. Pick two of these attributes. How do you think the attribute has changed over the years? Why?

   A. Average height of U.S. adult males (past 200 years)

   B. Cost of a personal computer (past 30 years)

   C. Average number of children in U.S. families (past 300 years)

   D. Number of people using the Internet (past 30 years)

Working with Lines of Fit

OVERVIEW

The goal of this lesson is to introduce students to interpreting scatter plots in the context of the Hall of Fame data set. The focus shifts from looking at changes over time to analyzing the relationship between two attributes. Some students have difficulty seeing an overall pattern in the points, especially because the points rarely form a perfect line. To address this challenge, the lesson introduces informal ways of drawing lines of fit and provides guiding questions to help students analyze the data.

Objectives

- Become familiar with the data set and the attributes
- Interpret scatter plots to determine the relationship between attributes
- Use informal methods to draw a line of fit
- Learn about different relationships between attributes: positive, negative, and no association

Offline

Class Time: One class period. The optional worksheet requires additional class time, or could be assigned for homework.

Materials

- Minutes and Points worksheet (one per student)
- Types of Associations reference sheet (one per student, on CD)
- Reading Scatter Plots worksheet (*optional,* on CD; use this worksheet if you want students to have more practice reading points on scatter plots)
- Minutes and Points Scatter Plot transparency (on CD)
- Rulers (one per student or pair)
- Chart paper (*optional*)

LESSON PLAN

Introduction

One way to brainstorm is to have students work in small groups. The groups get two minutes to brainstorm a list of attributes for basketball and write them on chart paper. When the time is up, post the lists and identify common attributes.

1. To set the context, have the class brainstorm a list of attributes that could be collected for basketball.
2. Introduce the data set, which has information on basketball players in the Basketball Hall of Fame. Students are likely to vary in their experience of and interest in basketball, so it's helpful to go over the meaning of some of the basketball terms used in the data set. Ask students to volunteer to explain basketball terms.
   - What is a field goal? A free throw? A rebound?

Exploration

3. Before giving out the worksheets, pose the question: Is there a relationship between the total number of minutes players played and the total number of points they scored? Students are likely to think that the more minutes players played the more points they are likely to have scored, especially because the players made it into the Hall of Fame. Explain that students will learn how to interpret a scatter plot to see if the data supports their hypothesis.
4. Display the Minutes and Points Scatter Plot transparency. This scatter plot shows the total number of minutes played and the total number of points for each player. Explain that the data are from each player's entire professional career. Ask a few questions to help students become familiar with the scatter plot.
   - What was the highest number of minutes played? How many points did that player score? [About 57,500 minutes; about 39,000 points]
   - What was the lowest number of minutes played? How many points did that player score? [About 7,500 minutes; about 4,000 points]
   - How many players both played for less than 15,000 minutes and scored less than 10,000 points? [3 players]
   - Let's get a sense of how much time these players spent playing basketball in their professional careers. How many hours is 57,500 minutes? About how many games (assuming 2 hours per game)? How many hours is 7,500 minutes? How many games? [57,500 minutes is about 958 hours, or about 476 games. 7,500 minutes is about 125 hours, or about 63 games.]

These questions are to help students get a sense of the amount of time because the numbers are so large.

Note: The scatter plot has 28 basketball players, instead of the 75 in the data set, to make it easier for students to read. In the next two lessons, students will work with the larger data set using TinkerPlots.

5. Now students need to shift from identifying individual points to looking for trends across the whole group of points. Some students have difficulty seeing an overall pattern in the scatter plot—they may view the plot as a jumble of random points floating in space. Explain that drawing a line of fit is helpful for seeing the overall trend of the data. Demonstrate how to draw a line of fit on the transparency. Then ask students to use rulers to draw a line of fit on their scatter plot.
6. Introduce the term *positive association*. Point out that the line of fit slopes upward from left to right. This indicates that there is a positive association between the two attributes: as one attribute increases the other increases. Explain that this is the overall trend of the data—for the group of players, playing more minutes tends to be associated with scoring more points. There may be individual players that are exceptions, but our focus is on the overall trend for the group. Point out that when students analyze scatter plots, they need to look at the "direction" of the line (going from left to right) to help them figure out if the association is positive or negative. Show examples of different associations by using the Types of Associations reference sheet.
7. Then explain that students also need to look at the "strength" of the association. Point out examples of weak and strong associations on the reference sheet. If the association is strong then you can use the line of fit to make predictions, because the predicted points are likely to fall on or close to the line. If the association is weak, then it wouldn't make sense to make predictions.
8. Have students work individually or in pairs on questions 5–7. Then, have a class discussion about the answers.

Wrap-Up

9. Have a class discussion about students' understanding of the relationships between attributes. Ask students: What kind of relationship do you think each pair of attributes is likely to have (positive, negative, or no association)? Encourage students to think generally about the relationship between the two attributes and not to focus on exceptions.

- Time spent reading and number of pages read [Positive]
- Time spent listening to music and number of songs heard [Positive]
- Number of letters in name and test scores [None]
- Age of a used car and selling price [Negative (until it becomes an antique)]
- Temperature and amount of money spent to heat home [Negative]
- Length of arm span and height [Positive]
- Height of a building and number of floors in building [Positive]
- Amount of change in pocket and number of pets [None]
- Come up with your own pair of attributes for each type of association.

ANSWERS

Minutes and Points

1. Most students will guess that as the number of minutes played increases, so does the number of points scored.
2. Sample answer:

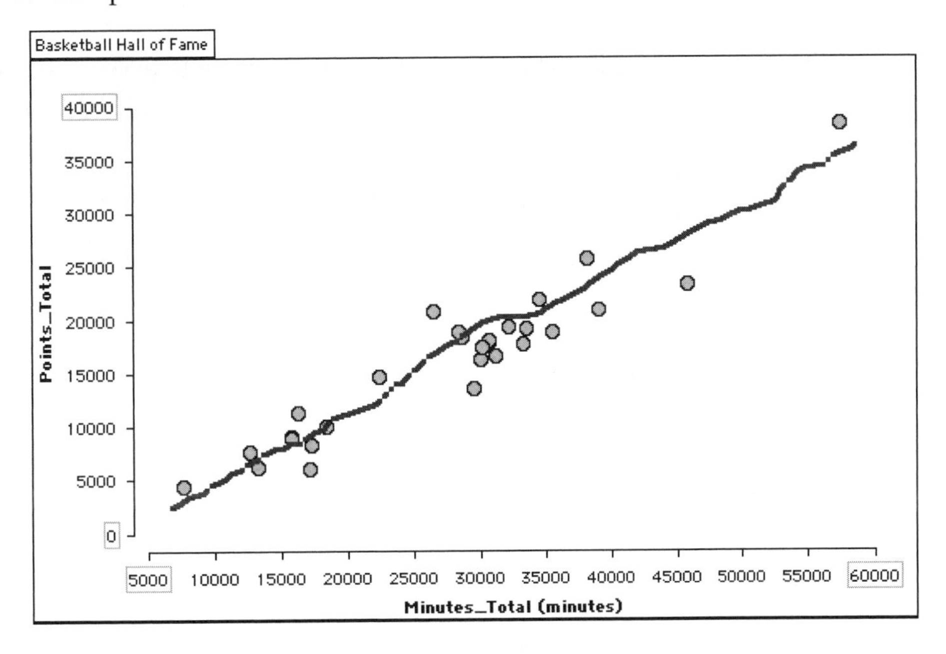

3. The line of fit slopes upwards.
4. a. Increase
   b. Lower
   c. Up; up (or down; down)
   d. Sample answer: The more minutes that a player played, the more points he scored.
5. The association is strong.

Reading Scatter Plots

1. J
2. K
3. I
4. J
5. B
6. A, B, C, D, E, F, G
7. A, B, C, F
8. G, H, I, J, K
9. There is a positive association between the two attributes, so as the number of games played increases, so does the number of points scored. The association is not very strong: it would be hard to predict a player's number of points based on his number of games.

Minutes and Points

Name:

You will analyze scatter plots to determine the relationship between two attributes.

ASK A QUESTION AND MAKE A HYPOTHESIS

You will be analyzing data on 28 basketball players from the Hall of Fame. Your task is to investigate the question: What is the relationship between the total number of minutes basketball players played in their professional careers and the total number of points they scored?

1. What do you think the data will show? Why?

ANALYZE DATA

To help you analyze the scatter plot, you will first learn how to draw a *line of fit*. Drawing a line of fit is a useful method for figuring out whether there is a relationship between two attributes.

Use a ruler and pencil to make a line that follows the trend of the data. In some scatter plots, most of the points will be on the line or close to the line. If the points are not close to a line, draw the line so that about half the points are above the line and half are below.

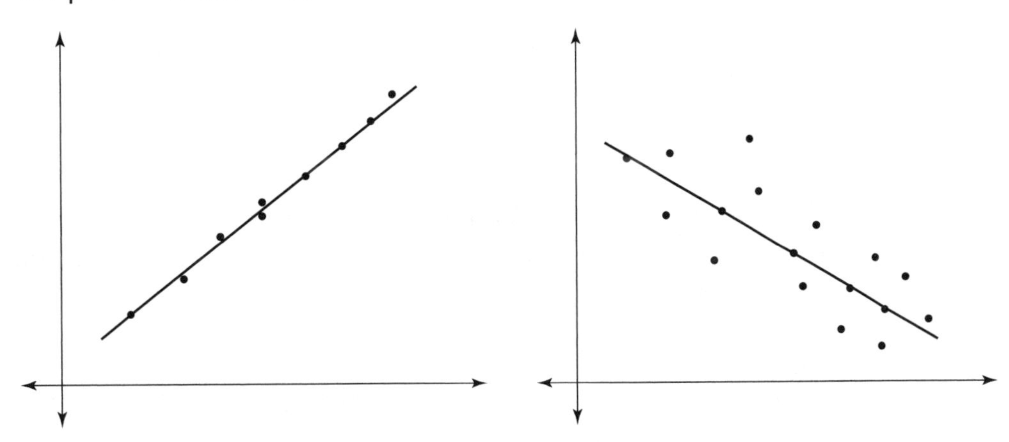

2. Draw a line of fit on the basketball scatter plot.

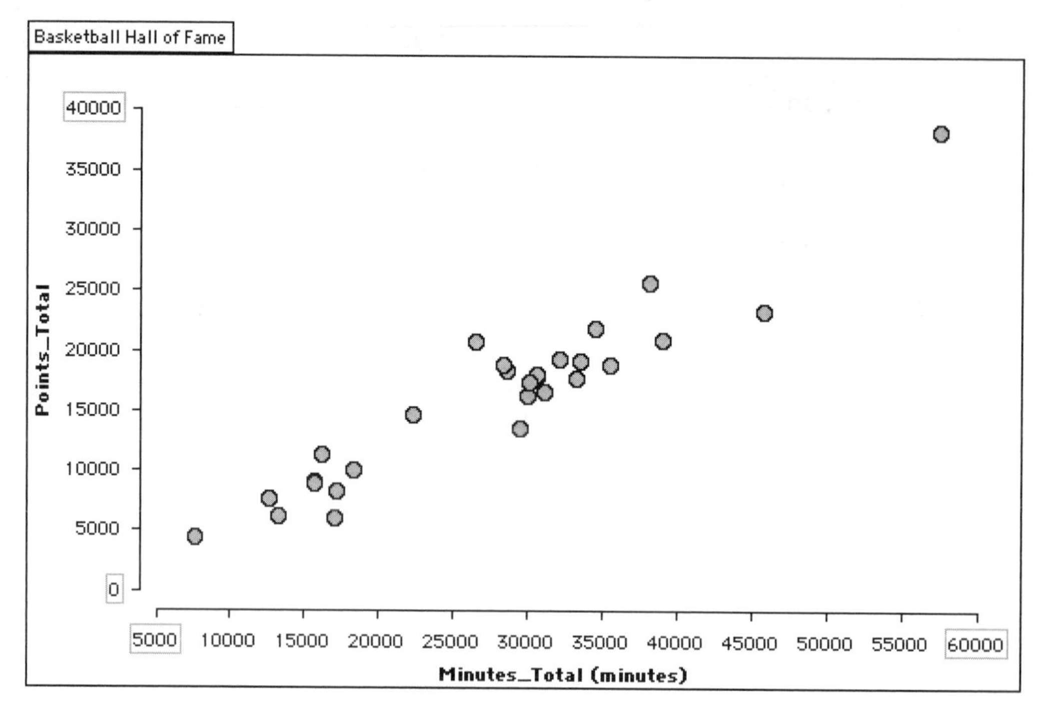

3. Describe the *direction* of the line of fit. Does it slope upwards or downwards from left to right? ____________________

4. The relationship between these two attributes is called a *positive association* because as the values for one attribute increase, the values for the other attribute increase. Fill in the blanks.

   a. When the number of minutes played increases, the number of points scored tends to __________.

   b. Players who played lower numbers of minutes tended to score __________ numbers of points.

   c. As the number of minutes played goes __________, the number of points scored tends to go __________.

   d. The more ____.

5. What is the strength of the association between these two attributes: minutes played and points scored? __________ (strong, weak, none)

Investigating Relationships Using Scatter Plots

OVERVIEW

The main focus of this lesson is on helping students to explore the relationship between two attributes. Students start with line plots because they have a lot of experience with this representation. They investigate the question: Is there a relationship between players' heights and the number of rebounds per game they make? They first display dots for the heights and color them with the number of rebounds per game. This enables students to see if the dots for higher heights also have darker colors, indicating more rebounds per game. Building on this work with the line plots, students use scatter plots to examine the relationship between pairs of attributes. They also compare scatter plots to determine which pair of attributes has a stronger relationship. Comparing scatter plots helps students to analyze the relationships between the attributes, because the contrast helps students to see the overall patterns in the points. (This lesson is based on an exploration by Clifford Konold and is used with permission.)

Objectives

- Explore the relationship between two attributes
- Create, interpret, and compare scatter plots

TinkerPlots

Class Time: One class period

Materials

- Heights and Rebounds worksheet (one per student)

Data Set: **Hall of Fame.tp** (75 basketball players from the Hall of Fame)

TinkerPlots Prerequisites: Students should be familiar with basic graphing, and using reference lines and dividers.

TinkerPlots Skills: Creating scatter plots and using color to display two attributes on a line plot are explained in this lesson.

LESSON PLAN

Introduction

1. Introduce the lesson goals and the first task: investigating the heights of the basketball players in the Hall of Fame.

2. Have students work individually or in pairs to analyze the data on the basketball players' heights. This activity helps students to become familiar with the data set and to use familiar graphs such as line plots before working with scatter plots.

3. When students have completed questions 1 and 2, have a short class discussion about their findings.

   - What did you find out about the heights for the group of basketball players?
   - What are typical heights?
   - What kind of plots did you make to investigate the heights?

4. Students next analyze line plots of the attribute *Rebounds_per_Game.* Briefly explain, or have a student explain, what *rebound* means in this context: The player catches a missed shot as it rebounds off the backboard. Have students work individually or in pairs to analyze the data on the players' numbers of rebounds per game. Then have a short discussion about their findings.

   - What did you find out about the number of rebounds per game for this group of basketball players?
   - What are typical numbers of rebounds per game?

5. Ask students to make their hypotheses (question 5): What is the relationship between players' heights and their number of rebounds per game? Do taller players tend to have more rebounds? What is the relationship between players' heights and points scored?

6. If you have a computer projection system, demonstrate how to use color in TinkerPlots for question 6. Make a line plot of heights. Then select the attribute *Rebounds_per_Game* in the data card but do not drag it to the axis. The dots will be colored by their values for *Rebounds_per_Game.* (The positions of the dots still represent the height values.)

   Ask students questions to build their understanding of the representation.

   - What does it mean if a dot has a dark color?
   - Where do the dark colors fall on the line plot—are there more dark dots in the shorter or taller heights?
   - Where do the light colors fall?

Exploration

7. Have students create the line plot on their own and analyze the relationship between the two attributes.

Let students know they can use dividers and reference lines to analyze scatter plots.

8. Have students work individually or in pairs to create and interpret their own scatter plots of the attributes *Height* and *Rebounds_per_Game.* They need to answer the questions about the form, direction, and strength of the scatter plot.

Wrap-Up

9. Have a class discussion about the investigation.
   - How would you describe the relationship in the scatter plot of *Height* and *Rebounds_per_Game?* Why do you think this is the case? [There is a weak positive association between height and rebounds per game, as shown by the slight upward slope of the dots.]
   - If the attributes have a strong association, does that mean that one attribute *causes* the other? [No, we cannot conclude that being taller *causes* players to have more rebounds per game.] It's important for students to recognize the difference between association and causation: Two attributes may have a strong association, but that does not mean that there is a cause-and-effect relationship between them.
   - If you knew a player's height, could you accurately predict the number of *Rebounds_per_Game* he scored? Why or why not?
   - How do these scatter plots compare with the ones for total minutes played and points scored in Lesson 5.3? Why did those scatter plots show a stronger association between the attributes?
   - Which player had the most rebounds per game for his height? [W. Russell is 82 inches tall and had 22.5 rebounds per game. Students can create plots similar to the ones they used to determine which player had the most points per game for his height.]

ANSWERS

Heights and Rebounds

2. Sample answer: The distribution has a wide center clump in the middle of the plot. The ends of the clump are taller than the center. Most basketball players have heights between 73 and 82 in. The most

typical heights are 76–80 in.: 44% of players have these heights. The median height is 78 in.

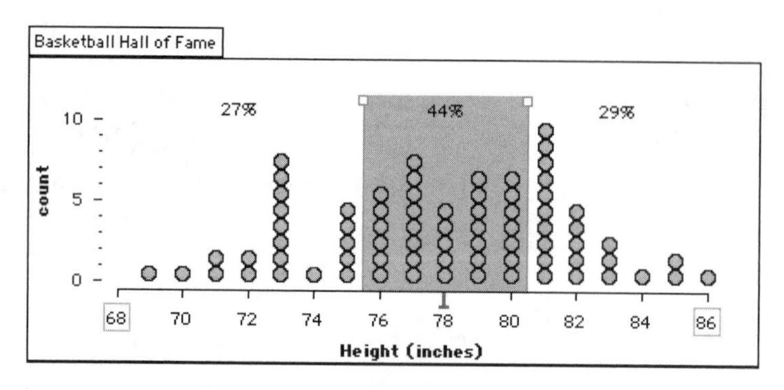

3. 6 rebounds per game

4. Sample answer: The distribution doesn't really have a center clump. It is flat, though it does tail off at the right end. So players are equally likely to have any number of rebounds per game between about 2 and 12. About 49% of players have between 4.1 and 10.1 rebounds per game. The median number of rebounds per game is 7.

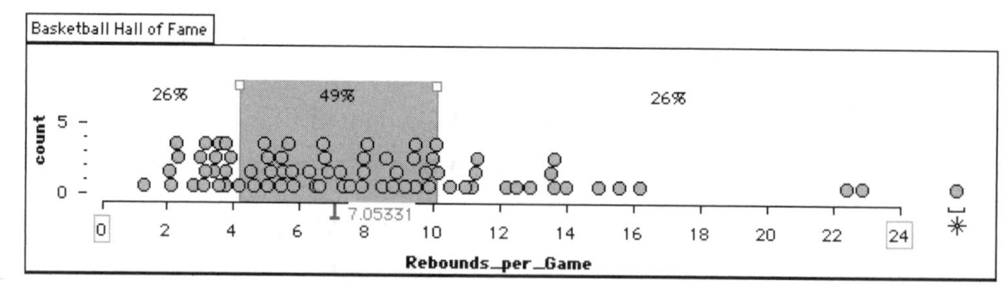

5. Sample answer: Taller players tend to get more rebounds because they can jump higher and get the ball closer to where it hit the backboard.

7. a. The dots are darker at 77 in. and above; most of the lighter dots are at 75 in. or below. The very dark dots are at 82 in. and above.

b. There is a positive relationship: The taller players do get more rebounds per game. Also, although no short players get a lot of rebounds, within the group of taller players, there is a lot of variation in the number of rebounds.

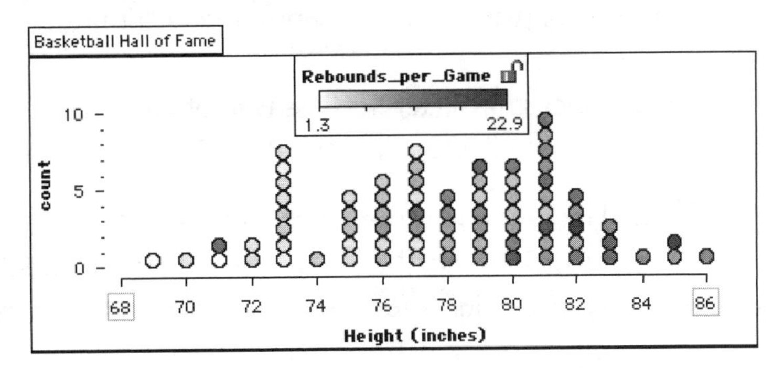

9. *Form:* The dots form a cone with the point at the bottom left and expanding up and to the right. *Direction:* The points slope upwards, meaning that taller players get more rebounds. *Strength:* The points don't lie on a straight line. There is a weak association. Taller players get more rebounds, generally, but you can't predict how many more.

10. Sample answer: There is a weak positive relationship between height and rebounds per game. Taller players do tend to get more rebounds per game, but their numbers of rebounds per game are more spread out than the shorter players. The taller a player is, the less accurately you can predict his number of rebounds per game.

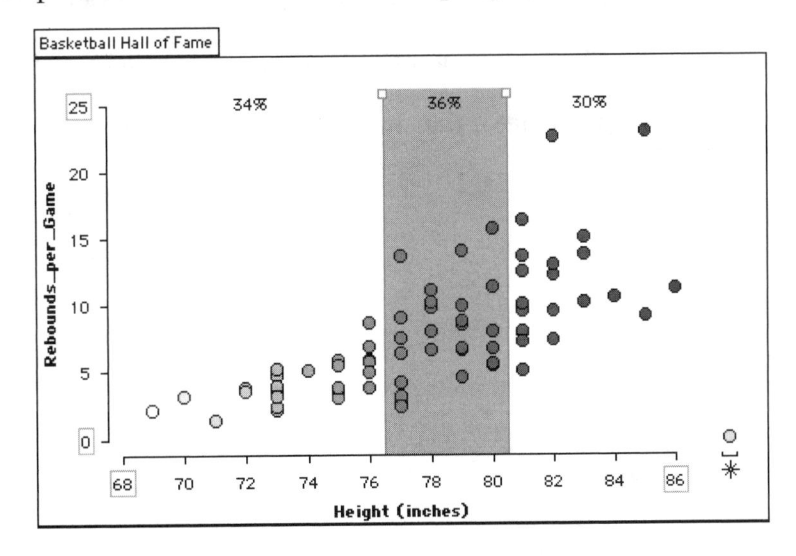

Heights and Rebounds

Name:

You will use scatter plots to investigate the relationships between basketball players' heights and other attributes.

1. Open the TinkerPlots file **Hall of Fame.tp** to see data about 75 basketball players in the Basketball Hall of Fame.

2. Find out about the heights of this group of players. Make a line plot of the heights. What is the shape of the distribution? What are typical heights for this group of basketball players? (Use dividers and percentages in TinkerPlots.)

3. A *rebound* is when a player catches a missed shot as it rebounds off the backboard. The attribute *Rebounds_per_Game* tells you the average number of rebounds a player had per game.

   If you played 10 games and had 60 rebounds, how many rebounds per game did you have? ________ rebounds per game

4. Make a line plot of the attribute *Rebounds_per_Game.*

   a. What is the shape of the distribution?

   b. What are typical *Rebounds_per_Game* for this group of basketball players?

ASK A QUESTION AND MAKE A HYPOTHESIS

5. Next you will investigate the question: What is the relationship between players' heights and their numbers of rebounds per game? For example, do taller players tend to get more rebounds per game? Make a hypothesis. What do you think the data will show? Why?

ANALYZE DATA

First, you will use color to look at two attributes on a line plot.

6. Make a line plot of players' heights. Then select *Rebounds_per_Game* in the data cards, but don't drag it to the axis. This will color the dots by their number of rebounds per game—the darker the color the higher the number of rebounds.
7. Use the colors to analyze the relationship between the attributes.
   a. Where are the darker dots on the line plot? The lighter dots?

   b. What does the color of the dots tell you about the relationship between rebounds per game and heights?

Next you'll make a scatter plot to explore the relationship between the two attributes.

8. Use your line plot of *Height.* Put *Rebounds_per_Game* on the vertical axis and fully separate the data vertically to make a scatter plot.

9. Complete the table. See the word bank for example words to use for describing scatter plots.

Questions about scatter plots	**Answers for the scatter plot of *Height* and *Rebounds_per_Game***
Form: Do the points seem to form an overall shape, such as a line or a curve, or are the points widely scattered across the plot?	
Direction: Do the points slope upwards or downwards from left to right? What does that show about the association between the two attributes?	
Strength: Do the points form a straight line or lie close to a line, or are they widely scattered? If you knew a player's height, could you accurately predict his rebounds per game?	

Word Bank

Form: line curve cluster cloud circle oval ellipse cone
Direction: upwards downwards flat positive negative association
Strength: strong weak tight loose

COMMUNICATE CONCLUSIONS

10. On a separate sheet of paper, write your conclusion for the question: What is the relationship between players' heights and their numbers of rebounds per game? For example, do taller players tend to get more rebounds per game?

    Make sure to write about the group of players and to use evidence from the data.

Connecting Line Plots and Scatter Plots

OVERVIEW

This lesson helps to deepen students' understanding of scatter plots by having students connect the position of points on line plots with their position on scatter plots. Students identify the middle 50% of the values on a line plot and then determine where these values fall on the scatter plot. This helps students understand what the positions of points mean on scatter plots, such as the location of points that represent players who were more successful at making free throws than was typical for the group. The scatter plots show a relationship between two attributes that are represented by a ratio (*Points_per_Game*) on one line plot and a percentage (*Free_Throw_Percent*) on the other. By connecting the two representations, students find out how a value on the line plot is represented by the relationship between two values on the scatter plot.

Objectives

- Recognize that each point on a scatter plot represents two values
- Interpret the positions of points on scatter plots
- Apply and deepen knowledge of ratios and percentages

TinkerPlots

Class Time: One to two class periods

Materials

- Scoring Points worksheet (one per student)
- Free Throws and Field Goals worksheet (one per student)

Data Set: Hall of Fame.tp (75 basketball players from the Hall of Fame)

TinkerPlots Prerequisites: Students should be familiar with intermediate graphing.

TinkerPlots Skills: Using highlighting to compare the position of points on two graphs is explained in this lesson.

LESSON PLAN

To highlight a group of points, draw a selection rectangle around them.

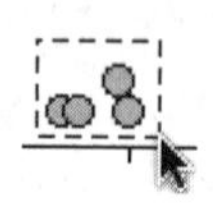

Introduction

1. Introduce the goal of the lesson, which is to continue exploring the data set for the basketball players in the Hall of Fame. Go over the directions for the activity. Students need to make a line plot of *Points_per_Game* and a scatter plot of *Games_Total* and *Points_Total.* If you have a computer projection system, demonstrate how to highlight points on the line plot to see where they fall on the scatter plot. Encourage students to first guess where they think the points will be on the scatter plot and to then use TinkerPlots to see if they are correct.

Exploration

2. Have students work individually or in pairs to find different points on the two plots for questions 6–9.

3. Go over the answers to the questions with the class. If time is short, focus on questions 6–9. Make sure students understand how the line plot is related to the scatter plot.

   - How did you figure out where to put the points in question 8?
   - How can you figure out a basketball player's *Points_per_Game* ratio by using the scatter plot only?

4. Introduce students to the next activity: Free Throws and Field Goals. This investigation is similar to the one they've just completed. Before they begin, you might want to review the definitions of *field goal* and *free throw* by having students explain the terms. Have students work individually or in pairs on the investigation. Students need to compare the scatter plot for free throws with the one for field goals.

Wrap-Up

5. Have a class discussion about the Free Throws and Field Goals investigation.

   - Where did you add a point to the scatter plot for a player who is more successful at making free throws than most of the other players? [You can add a point where the number of free throws made is close to the number of free throws attempted, such as 8000 attempts and 7500 made. This new point would be positioned above the line of points.]

Make sure students realize that the number of free throws made can not be greater than the number of free throws attempted.

- Where did you add a point to the scatter plot for a player who is less successful at making free throws than most of the other players? [You can add a point where the number of free throws made is much lower than the number of free throws attempted, such as 8000 attempts and 2000 made. This new point would be positioned below the line of points. The lowest possible number of free throws made is 0 but that would be very unlikely for this group of players.]
- How does the information you get from the scatter plot of *Free_Throw_Attempts* and *Free_Throws_Made* compare with the information in the line plot of *Free_Throw_Percent*?
- Are these players more successful with free throws or field goals? How can you tell by comparing the scatter plots? [In both scatter plots, the points form lines that slope upwards. The slope is higher for the free throws than for field goals, which shows that players made a higher percentage of free throws than of field goals. Students can check this by comparing the line plots of *Free_Throw_Percent* and *Field_Goal_Percent*.]
- In this lesson we saw a relationship between the positions of points on a line plot and their positions on a scatter plot. Would any line plot and any scatter plot have this relationship? [This relationship happens only when the attribute in the line plot is the ratio of the two attributes in the scatter plot.]
- What other questions would you like to investigate about basketball players? How would you collect and analyze the data?

ANSWERS

Scoring Points

2. b. 9.8 points per game

3. The middle 50% of the players scored 14.6–20.8 points per game. The median is 18.5 points per game.

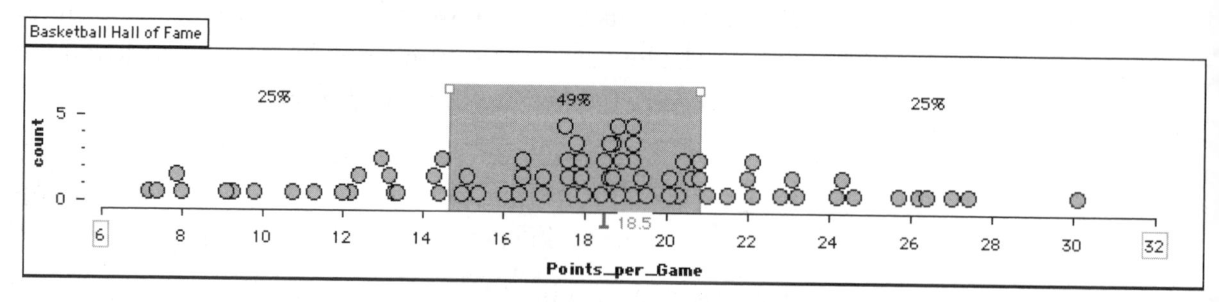

5. See the worksheet for an example.

6. a. Guesses will vary.
   b. Near the line of fit

7. a. Students may guess above the line of fit.
   b. Above the line of fit

8. a. Player H will be above the line of fit and at the right end of the line plot.
   b. Player L will be below the line of fit and at the left end of the line plot.
   c. Player T will be on the line of fit and in the middle of the line plot.

9. W. Chamberlain, E. Baylor, and J. West. These players have the three highest *Points_per_Game* ratios, so they are the rightmost points on the line plot. On the scatter plot, these players are the farthest above the line of fit. (*Note:* If these players are not the farthest above the line of fit for some students, they should consider moving their line.)

Free Throws and Field Goals

2. a. 75%
   b. 6 players

c. The middle 50% of players have free throw percents between 75.0% and 80.6%. The median free throw percent is about 77.7%.

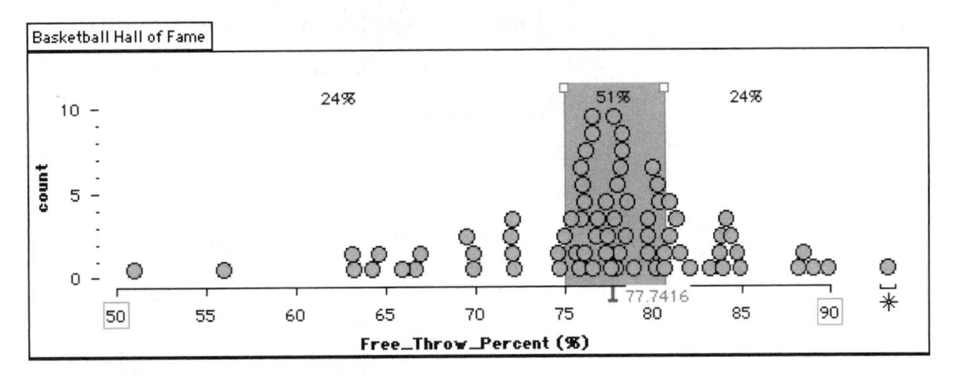

4. Sample answer:

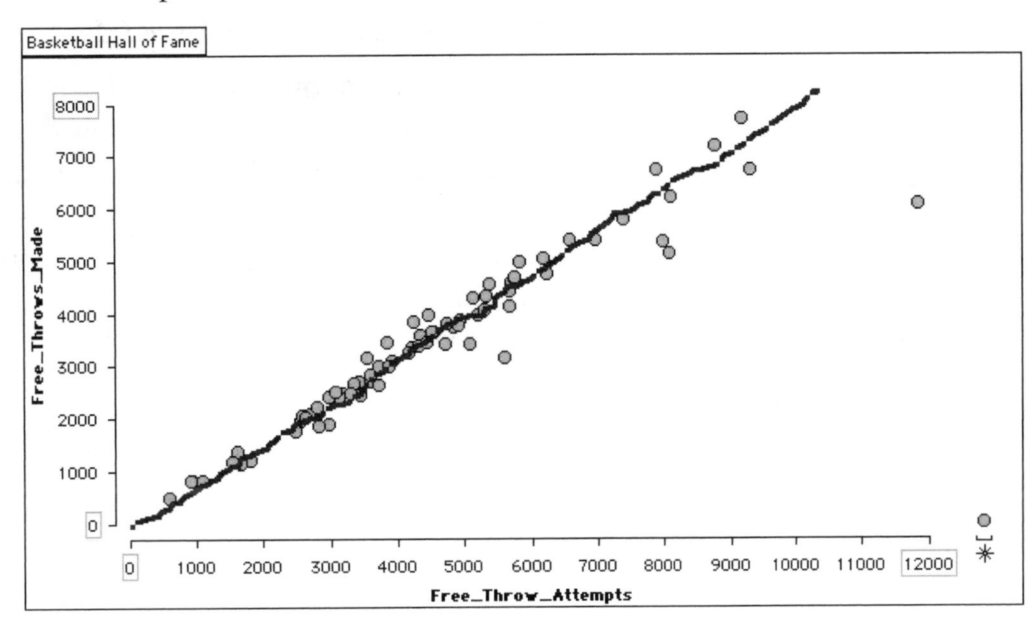

5. a. The middle 50% fall near the line of fit.
   b. The bottom 25% fall below the line of fit.
6. Sample answer: Players tend to be more successful with free throws; the line of points for free throws is steeper than the line for field goals.

7. Sample answer: The points in the free throw graph form a steeper line than the field goal points. The free throw percent is 75%, so I think the field goal percent would be less than that. Using TinkerPlots, a player typically makes fewer than half the field goals he attempts (42%–50%). The median field goal percent is 44.8%.

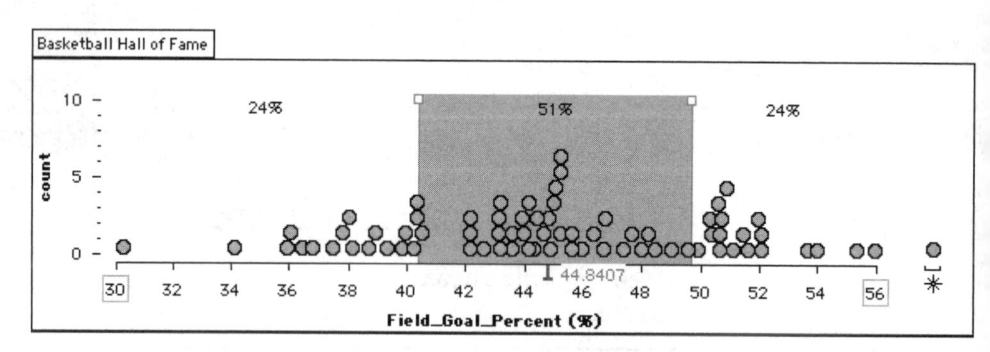

Note: Some students might apply the highlighting technique of this lesson to the two line plots to see whether players who have high free throw percents also have high field goal percents. This is an interesting extension, and in fact, there does not seem to be a relationship between field goal percents and free throw percents. Dots in the highest 25% of one graph appear in all parts of the other graph, and so on.

Scoring Points

Name:

You will use two different plots to analyze the relationship between games played and points scored for basketball players in the Hall of Fame.

1. Open the TinkerPlots file **Hall of Fame.tp** to see the data on 75 players who are in the Basketball Hall of Fame.

2. The attribute *Points_per_Game* is the average number of points the player scored per game. It is the ratio of *Points_Total* to *Games_Total.*

$$Points\_per\_Game = \frac{Points\_Total}{Games\_Total}$$

   For example, if you scored 200 total points and had played 10 games, you would have scored 20 points per game.

   a. Make a line plot of *Points_per_Game*. Find the dot for W. Chamberlain, who has a *Points_per_Game* ratio of 30.1.

   b. Find the dot for S. Martin. He played a total of 745 games and scored a total of 7337 points. What is his *Points_per_Game* ratio?

   ____

3. Use the dividers and percentages to find the middle 50% of the data. What are typical values for *Points_per_Game* for this group of players?

   ____

4. On a new plot, make a scatter plot with *Games_Total* on the horizontal axis and *Points_Total* on the vertical axis.

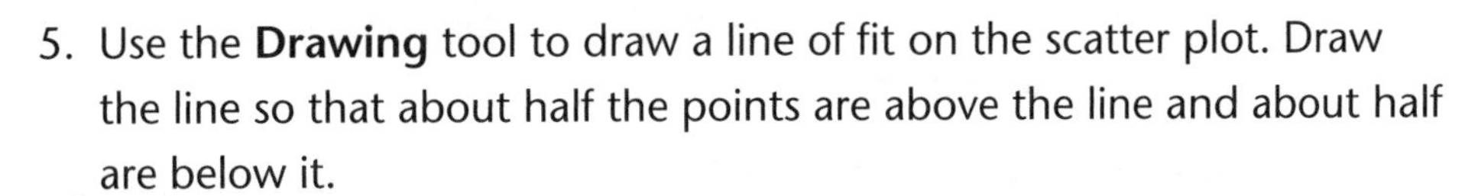

5. Use the **Drawing** tool to draw a line of fit on the scatter plot. Draw the line so that about half the points are above the line and about half are below it.

For the next two questions, you will select points in the line plot and see where they appear in the scatter plot. For example, if you select the four lowest dots on the line plot, TinkerPlots will automatically highlight those points on the scatter plot.

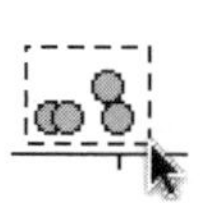

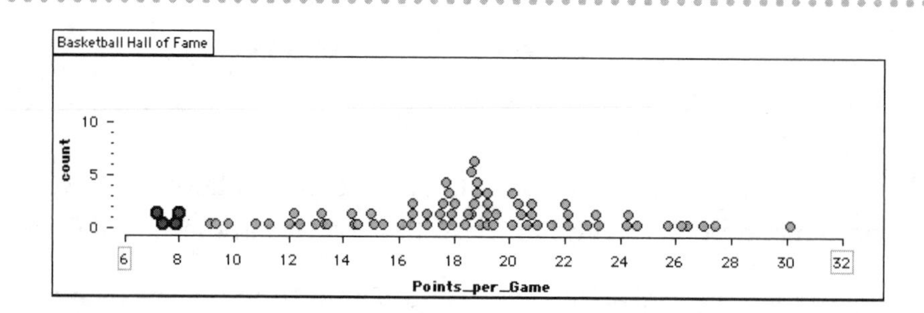

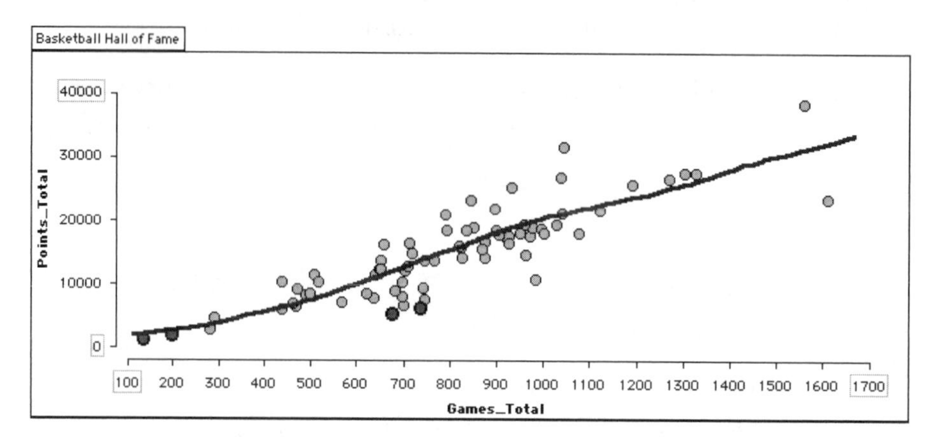

6. Where do you think the middle 50% of the *Points_per_Game* values will appear on the scatter plot?

   a. Guess first. What is your guess? ______________________

   b. Select all the dots in the middle 50% on the line plot. Where are those points on the scatter plot? ______________________

7. Where do you think the top 25% of the *Points_per_Game* values will appear on the scatter plot?

   a. What is your guess? ______________________

   b. Highlight the top 25% on the line plot. Where are the points on the scatter plot? ______________________

8. Add each new point to both plots above and label the points.

   a. Player H has a higher *Points_per_Game* ratio than most players.

   b. Player L has a lower *Points_per_Game* ratio than most players.

   c. Player T has a typical *Points_per_Game* ratio.

9. Which three basketball players are the best at scoring points? Where are these players on each plot?

Free Throws and Field Goals

Name:

You will be investigating data on free throws and field goals in basketball. A *free throw* is a penalty shot.

- *Free_Throw_Attempts* is the number of free throws tried.
- *Free_Throws_Made* is the number of successful free throws.
- *Free_Throw_Percent* is the percentage of free throws that are successful.

A *field goal* is any successful shot, except a free throw, that is made during game play.

FREE THROWS

1. Open the TinkerPlots file **Hall of Fame.tp** to see the data on players who are in the Basketball Hall of Fame.
2. Make a line plot for the attribute *Free_Throw_Percent.*
   a. If you made 30 free throws out of the 40 free throws you attempted, what is your free throw percent (*Free_Throw_Percent*)?
   ____
   b. How many players have a *Free_Throw_Percent* that is less than 65%? __________
   c. What are typical free throw percentages for this group of players? Use evidence from the data.
3. Make a scatter plot with the attribute *Free_Throw_Attempts* on the horizontal axis and *Free_Throws_Made* on the vertical axis.

4. Draw a line of fit on the scatter plot.

5. Experiment with highlighting points on the line plot to see where they appear on the scatter plot.

   a. Where do the middle 50% of the *Free_Throw_Percent* values appear on the scatter plot ? ____________________

   b. Where do the bottom 25% of the *Free_Throw_Percent* values appear on the scatter plot? ____________________

COMPARE FREE THROWS AND FIELD GOALS

Free throws

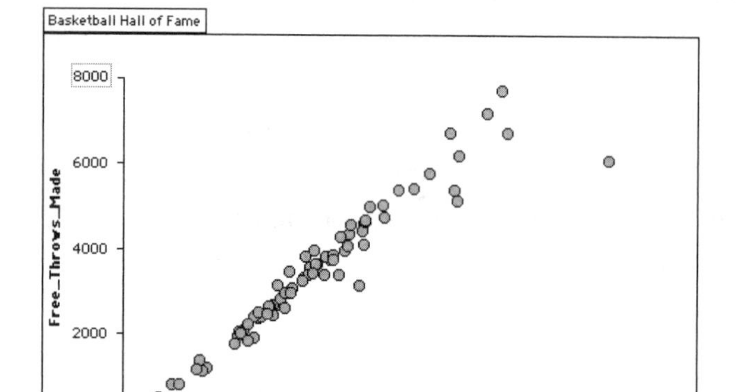

Field goals

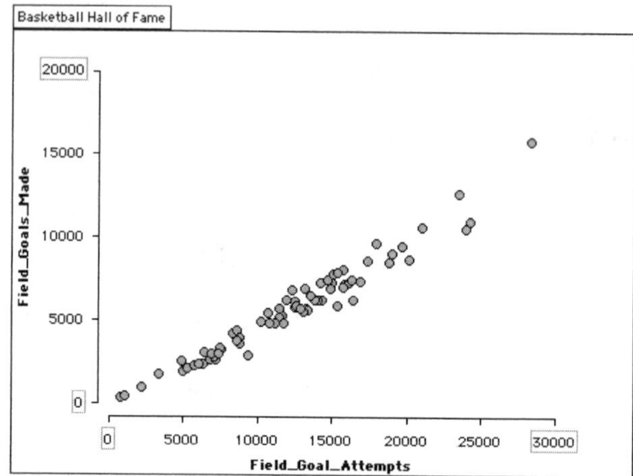

6. Does this group of players tend to be more successful with free throws or field goals? How can you tell from the scatter plots?

7. For this group of players, a typical *Free_Throw_Percent* value was between 75% and 81%. What are typical *Field_Goal_Percent* values for the group? Make a prediction and then use TinkerPlots to find out. Explain how you made your prediction.

Exploring Data from Basketball Teams

OVERVIEW

This lesson is a shift from the focus on scatter plots in the previous lessons. Students conduct a more open investigation of the question: Is there a home-court advantage for the Carolina Tar Heels basketball team? Students need to figure out which attributes to investigate, what kinds of plots to make, and what they would consider as evidence for a home-court advantage. They can approach this investigation with varying levels of complexity. As a starting point, students compare the percentage of home games won to the percentage of away games won. They then create a new attribute, *Diff_Score*, to determine whether the team won by more points at home games than at away games.

Objectives

- Determine which attributes to analyze to investigate a question
- Create a variety of plots to analyze data
- Use a formula to create a new attribute
- Write conclusions that use the data as evidence

TinkerPlots

Class Time: Two class periods

Materials

- Is There a Home-Court Advantage for the Tar Heels? worksheet (one per student)
- Is There a Home-Court Advantage for College Teams? worksheet (one per student)
- Lab Sheet for Home-Court Advantage (two per student)
- Rubric for Investigating Relationships (on CD)

Data Sets: Tar Heels.tp (data from the North Carolina Tar Heels, a women's basketball team). In the second activity, students can choose different teams from the files in the **Extra Basketball Data** folder.

TinkerPlots Prerequisites: Students should be familiar with intermediate graphing and creating a new attribute by making a formula.

TinkerPlots Skills: Using a filter is explained in this lesson.

LESSON PLAN

Introduction

1. Introduce the data set, which is from the Carolina Tar Heels, a women's basketball team from the University of North Carolina. The data include scores for 35 games from the 2001–2002 season and were collected from information on the team's website.

2. Introduce the question: Is there a home-court advantage? To make sure that students understand the question, have a short class discussion about the meaning of *home-court advantage.*

   - What does "home-court advantage" mean?
   - What would convince you that a team had (or did not have) a home-court advantage?
   - Which attributes will you investigate to determine whether the Carolina Tar Heels have a home-court advantage?

3. Have students write their hypotheses individually (question 1). Ask a few students to share their hypotheses with the class.

Exploration

4. Have students work independently or in pairs to analyze the data. Some questions to encourage student exploration and check for understanding:

   - Which attributes are you analyzing? Why?
   - Are there other attributes that would be helpful to analyze the data? If so, which?
   - What kind of evidence have you found to support or not support a home-court advantage?
   - What plot would you show someone to convince him or her that there is strong evidence for your conclusion?

As you circulate around the room, look for particularly convincing plots for students to share with the whole class in the wrap-up discussion.

5. In question 4, students need to analyze the attributes *Site* and *Results.* They should show percentages to look at how many games the Tar Heels won, and where. Tinkerplots automatically displays percentages for *rows* of data. Students may not realize this and thus might interpret the percentages incorrectly. You may want to do a whole class demonstration to show how the percentages work and how to change the kind of percentages that are displayed. The left graph shows *row*

To show column percentages, choose **Show Column Percent** from the menu to the right of the **%** button.

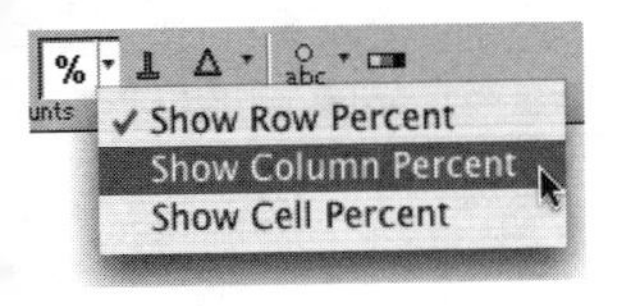

percentages: it shows that 58% of the games the Tar Heels won were played at home. The right graph shows *column percentages:* it shows that the Tar Heels won 83% of the games played at home. This is a subtle distinction. You could also have students show *cell percentages* to look at what percentages of the total number of games were won and at home.

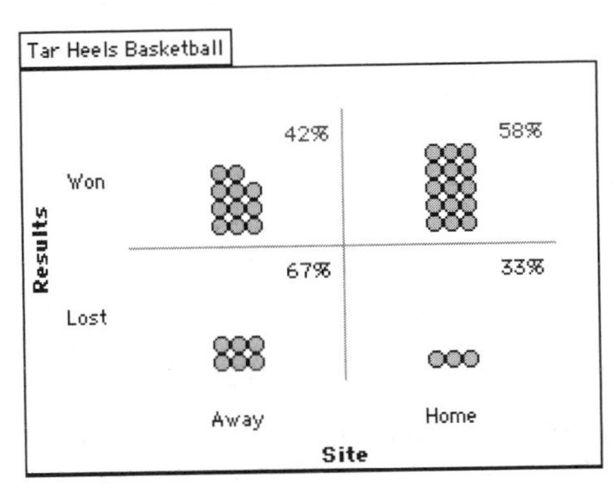

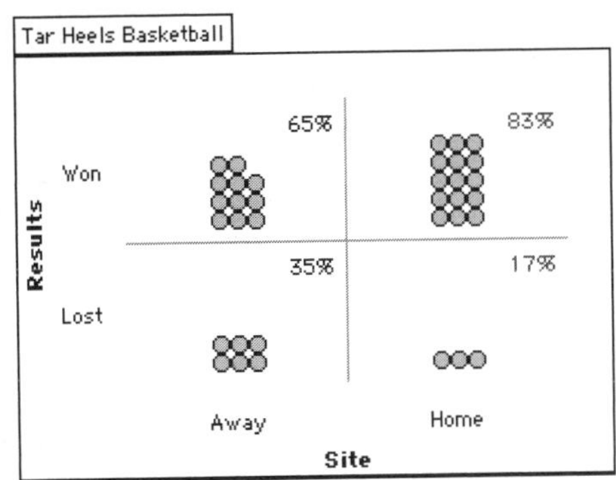

It's important for all the students to use the same formula so that the class will have common data to discuss.

6. The lesson suggests that students create a new attribute by using a formula with TinkerPlots (question 5). The attribute *Diff_Score* shows the difference in final score between the Tar Heels and their opponents. By analyzing this attribute, students can determine whether the Tar Heels tend to win by more points at home games than away games. Similarly, students can determine whether the team loses by fewer points at home than at away games. This lesson assumes that students have had experience creating formulas with TinkerPlots. If students are new to creating formulas, it's helpful to demonstrate making the formula *Diff_Score* = *Team_Score* – *Opponent_Score*.

   a. In the Attribute column of the data card, double-click **<new attribute>** and type `Diff_Score`.

   b. Expand the data cards until you see the Formula column. Double-click the circle for *Diff_Score*. A formula editor will appear.

   c. You can either type the formula or use the keypad. Expand the Attributes list to see the attribute names. Double-click a name to enter it in the formula.

   d. When you're done, click **OK**.

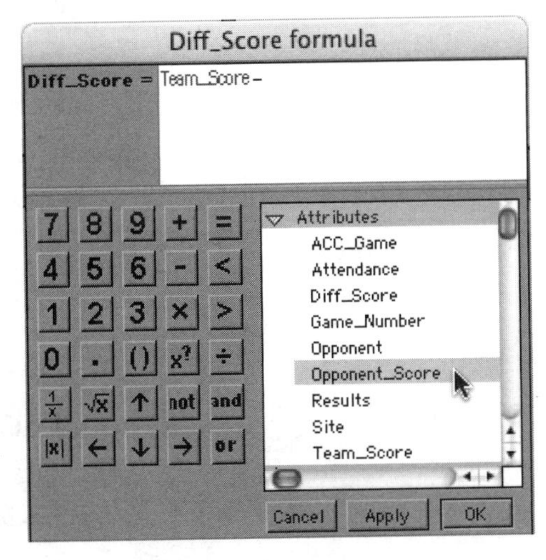

Here is a line plot of the attribute *Diff_Score*. There are more positive values than negative values, which shows that the Tar Heels won more games than they lost.

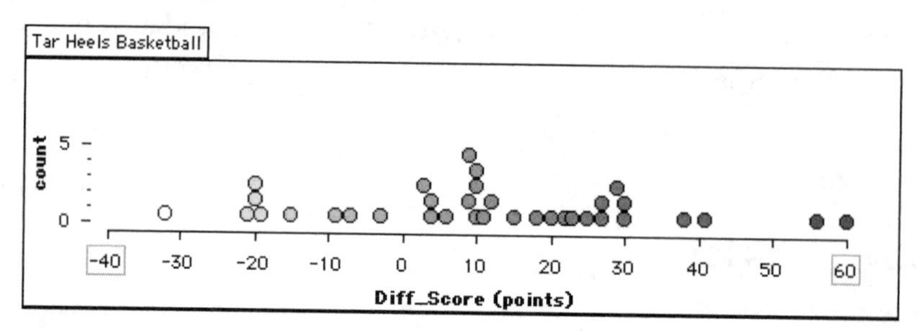

To see if students understand the new attribute, *Diff_Score*, you may want to ask questions as you circulate around the room:

- What does it mean when *Diff_Score* has a positive value, such as 25? [The Tar Heels won the game by 25 points.]
- What does it mean when *Diff_Score* has a negative value, such as –15? [The Tar Heels lost the game by 15 points.]
- Which dots represent the games that the Tar Heels won? [Positive values] Lost? [Negative values]. If students are confused, select *Results* to color the dots by Won or Lost.
- How did the difference in scores compare for the home games and the away games? What kind of plot can you make to investigate this question?

7. Students then use a filter to look at the data for only the games that the Tar Heels won. The directions are on the worksheet.

   Troubleshooting tips for filters:

   - When you create a filter, you enter an expression specifying a property for points that you want to appear in the graph. You do need to enter a complete expression. This is different from creating a new attribute, when the left side of the expression is entered for you.
   - Dividers won't change position when you filter. Make sure you have any dividers on the positive side of the graph before you filter (or turn them off and on again to fix the problem).
   - To get rid of the filter, go to the **Plot** menu and choose **Remove Filter.**

   Students could create a different filter, *Results* = "Lost", to investigate the data for the games lost. However, it is difficult to draw strong conclusions based on only nine games lost.

Wrap-Up

8. Have a class discussion about the results. Ask a few students to share their plots with the whole group.
   - What are your conclusions? What evidence did you find to support your conclusions?
   - What kinds of plots did you find to be particularly helpful for analyzing the data?
   - Do you think that winning games at home is enough evidence of a home-court advantage? Do you think it makes a difference how many points the team won by? For example, does winning by 2 points mean the same thing as winning by 22 points when considering a home-court advantage?
   - What did you find out from creating and analyzing the new attribute, *Diff_Score*?
   - Do you think you would get similar findings if you analyzed data for other basketball teams? For middle-school teams? For NBA teams? Why or why not?
   - The Tar Heels won 75% of their games. How would you look for evidence of a home-court advantage in a team that won 50% of its games?
9. If you want students to write their conclusions, go over the Rubric for Investigating Relationships. Point out that this rubric is similar to the ones that students have used for writing conclusions in other lessons.
10. Have students investigate the same question for different basketball teams. Students can choose data sets for ten basketball teams (five men's and five women's.) Students can also prepare data sets from their favorite teams by using information from teams' websites.
    - Choose a different basketball team to investigate. Does this team have a home-court advantage?
    - Compare your findings for the new team with those for the Carolina Tar Heels. How are the findings similar or different? Which is a stronger case for a home-court advantage?
11. As a wrap-up, have students prepare and give short presentations on their findings for the other basketball teams.

ANSWERS

Is There a Home-Court Advantage for the Tar Heels?

1. Sample answer: I think there will be a home-court advantage. Sports teams I've watched seem to have a home-court advantage.

3. Students should look at *Site, Results, Team_Score,* and *Opponent_Score.*

4. Sample answer: *Site* and *Results*: More games were won at home (83%) than lost at home (17%), however more games were won away (65%) than lost away (35%) as well.

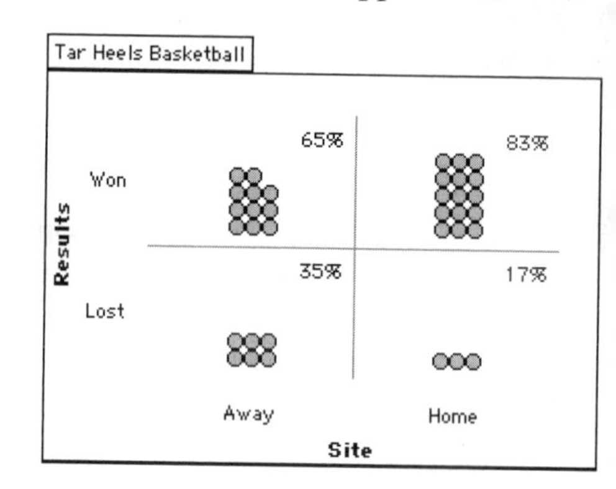

Opponent_Score and *Site:* Opponents tended to score more points, and to have more variation in their scores, during away games. At away games, opponents' scores were typically about 59–86 points, with a median score of 71 points. At home games, opponents' scores were typically about 58–69 points, with a median score of 65 points.

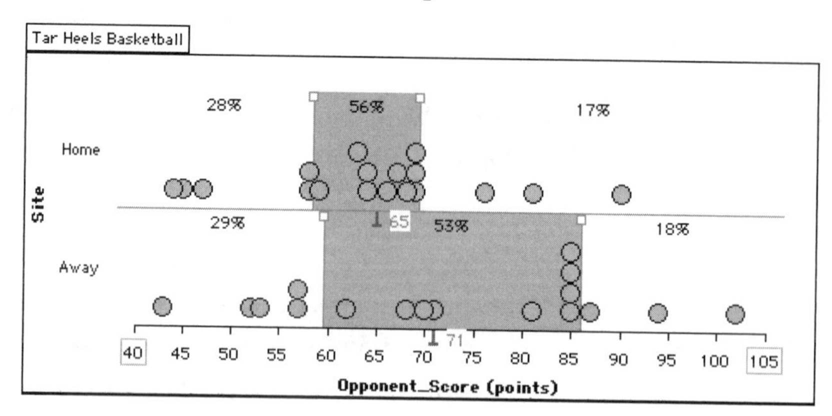

Team_Score and *Site:* The team's scores were usually better when they were at home. At home games, the team's scores were typically about 75–91 points, with a median score of 85 points. At away games, the team's scores were typically about 66–86 points, with a median score of 78 points.

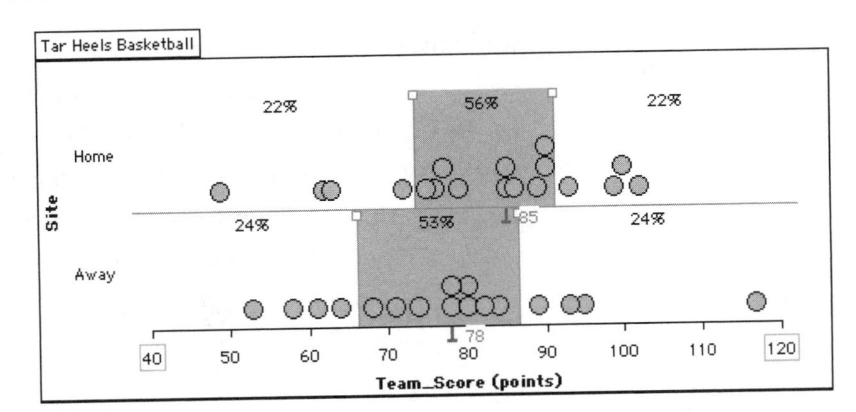

6. The difference in scores was generally larger for home games. At away games, the difference in scores was typically about –15 to 11 points, with a median difference of 9 points. At home games, the difference in scores was typically about 3 to 29 points, with a median difference of 21 points.

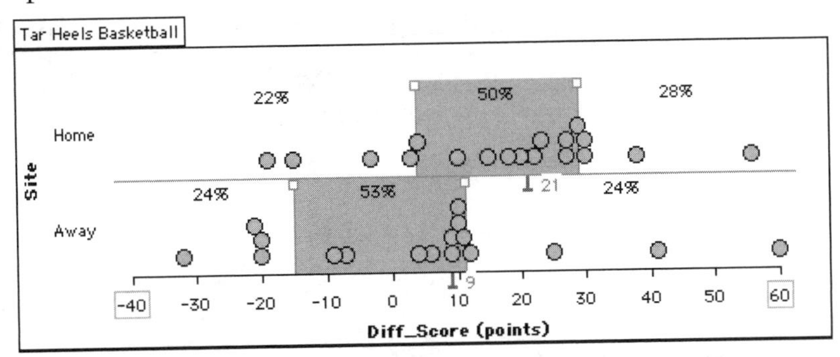

9. Looking at only the games the Tar Heels won, they did tend to win by more points at home games than at away games. This graph is divided roughly into thirds because the distribution of points makes finding a middle 50% difficult. At home games, the Tar Heels typically won by 19–22 points, with a median win by 23 points. At away games, they typically won by 9–18 points, with a median win by 10 points. So typically the Tar Heels won home games by 13 more points than they won away games.

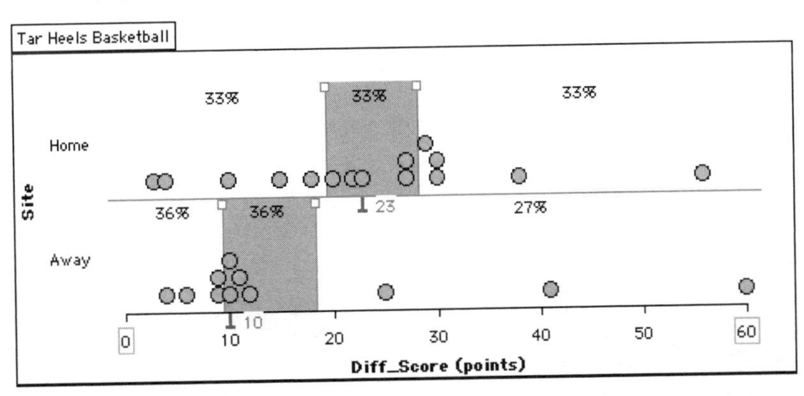

10. Students' conclusions should include their answers to questions 4–9. Sample answer: The data show support for a home-court advantage for the Carolina Tar Heels. The Tar Heels won 83% of the games they played at home compared to 65% of the games they played away. Of the games that the team won, 58% were played at home and 42% were played away. The Tar Heels won by a mean of 18 points at the home games compared to a mean of 5 points at the away games. The games were closer at the away games, and they had a bigger lead at the home games, which gives strong support for a home-court advantage. If you look at only the games the Tar Heels won, they tended to win by fewer points at the away games and by more points at the home games. 60% of the home games were won by 20–40 points, compared to only 9% of the away games. 73% of the away games were won by less than 20 points compared to 33% of the home games.

    In this grid, the percentages are given for the rows. 58% of the games the team won were at home and 42% were away.

    In this grid, the percentages are given for the columns. This grid shows that the Tar Heels won 83% of the games they played at home compared to 65% of the games that they played away.

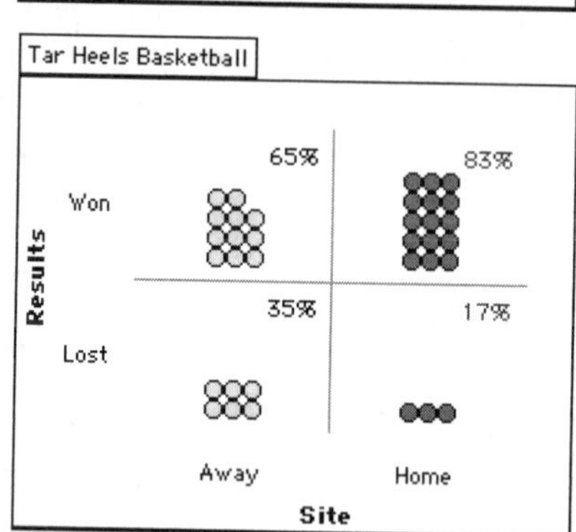

To see column percentages, choose **Show Column Percent** from the menu to the right of the **%** button.

This plot shows that the Tar Heels won by a mean of 18 points at the home games compared to a mean of 5 points at the away games. The games were closer at the away games, and they had a bigger lead at the home games. This supports a home-court advantage.

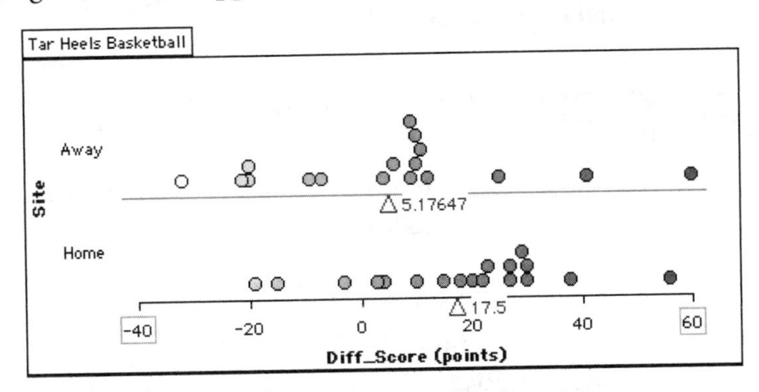

This plot uses a filter to show only the games that the Tar Heels won (27 games). It shows that 73% of the away games were won by less than 20 points compared to 33% of the home games. 60% of the home games were won by 20–40 points compared to only 9% of the away games.

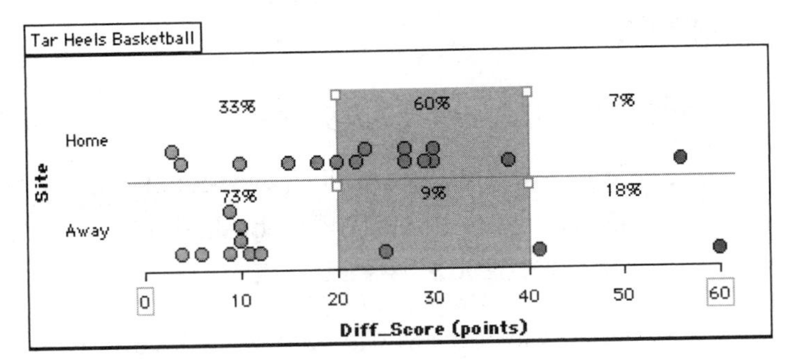

Is There a Home-Court Advantage for College Teams?

Answers will vary. These teams do show evidence of a home-court advantage: Arizona Men's, Arizona Women's, Boston College Men's, Boston College Women's, Connecticut Men's, Connecticut Women's, Maryland Men's, North Carolina Men's, Seton Hall Men's, and Stanford Men's. These teams do not show evidence of a home-court advantage: Duke Men's, Duke Women's, Maryland Women's, North Carolina Women's, Seton Hall Women's, and Stanford Women's.

Is There a Home-Court Advantage for the Tar Heels?

Name:

You will analyze data by creating a variety of plots and by using a formula to make a new attribute.

ASK A QUESTION AND MAKE A HYPOTHESIS

You will investigate the question: Is there is a home-court advantage for the Carolina Tar Heels basketball team? Does this team tend to do better when it play games on its home court than when it plays games away?

1. What do you think the data will show? Give reasons for your hypothesis.

ANALYZE DATA

Make Different Plots

2. Open the TinkerPlots file **Tar Heels.tp** to see the scores from 35 games for the women's basketball team, the Carolina Tar Heels, for the 2001–2002 season.

3. What attributes in the data set will you investigate to find out whether there is a home-court advantage? List the attributes that you plan to analyze.

___

4. Make at least three different plots to investigate the question: Is there a home-court advantage for the Carolina Tar Heels? (*Tip:* To add a new graph, drag the Plot icon from the toolbar.)

   - The graphs need to have different axes. Making a histogram and then changing the bin size to make a new histogram does not count as a different graph.
   - Do not make circle graphs for this activity.

   Take notes about your investigation in the table on the next page.

What attributes did you investigate?	What did you find out? What does the data show about whether or not there is a home-court advantage?
Site and *Results*	

Use a Formula to Find the Difference in Scores

5. Use TinkerPlots to make a formula to find the differences between the Tar Heels' scores and their opponents' scores. Call the attribute *Diff_Score*.

$$Diff\_Score = Team\_Score - Opponent\_Score$$

6. Make plots of the attribute *Diff_Score* to investigate the question: How does the difference in scores compare for home games and away games? Keep notes on your findings.

Use a Filter

You can use a *filter* to look at the data for only the games that the Tar Heels won. This will help you to answer the question: Did the Tar Heels tend to win by more points when they played home games compared to away games?

7. Select the plot (click on it). Go to the **Plot** menu and choose **Add Filter.**

8. Type `Results = "Won"`. Click **OK.** The plot will have 26 dots to show only the games that were won.

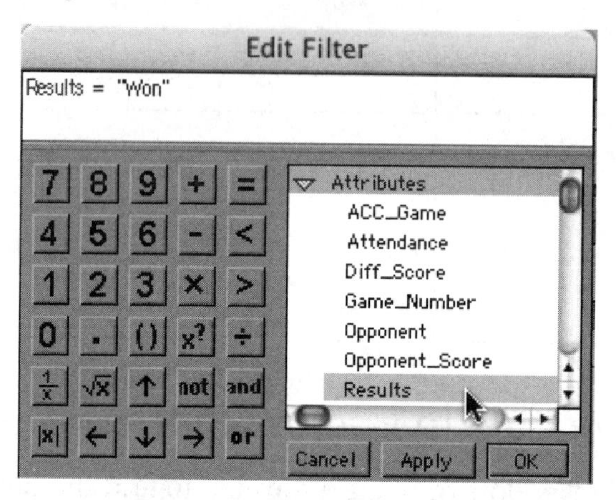

9. Analyze the data to answer the question: Did the Tar Heels tend to win by more points when they played home games compared to away games? How many more points? Write down your findings.

COMMUNICATE CONCLUSIONS

10. On a separate sheet of paper, write your conclusion for the question: Is there a home-court advantage for the Carolina Tar Heels?

Is There a Home-Court Advantage for College Teams?

Name:

You will continue your investigation of the question: Is there is a home-court advantage in college basketball? You looked at the data for one team, the Carolina Tar Heels. Now, you will investigate data from more college basketball teams.

ASK A QUESTION AND MAKE A HYPOTHESIS

1. Make a hypothesis. Do you think that there will be evidence of a home-court advantage for other teams? Why or why not?

2. Choose at least two different basketball teams to analyze. Circle your choices below.

   Women's NCAA: Arizona Boston College Connecticut Duke Maryland North Carolina Seton Hall Stanford

   Men's NCAA: Arizona Boston College Connecticut Duke Maryland North Carolina Seton Hall Stanford

3. How will you analyze the data to see if there is a home-court advantage? Think about the methods that you used for the Tar Heels.

   a. Which attributes will you investigate? What new attributes will you create? Why?

   b. What kinds of graphs will you make? Why?

4. Use your plan to analyze the data for each basketball team. Keep notes on your findings on the Lab Sheet.

Lab Sheet for Home-Court Advantage

Name:

Name of Basketball Team: ____________________

ANALYZE DATA

1. Use TinkerPlots to investigate the data for the team. Keep your notes in the table.

What evidence did you find in *support* of a home-court advantage?	What evidence did you find *against* a home-court advantage?

COMMUNICATE CONCLUSIONS

Prepare a short presentation to share your findings with the class.

2. Decide what graphs to show the class. Pick graphs that provide strong evidence for your conclusions. Add a text box to explain what each graph shows. Print your graphs.

3. Plan what you will say by writing responses to these questions on a separate sheet of paper. Make sure to give evidence from the data.

   a. Is there a home-court advantage for this basketball team? Why or why not?

   b. How do your findings for this basketball team compare with the findings for the Tar Heels and other basketball teams?

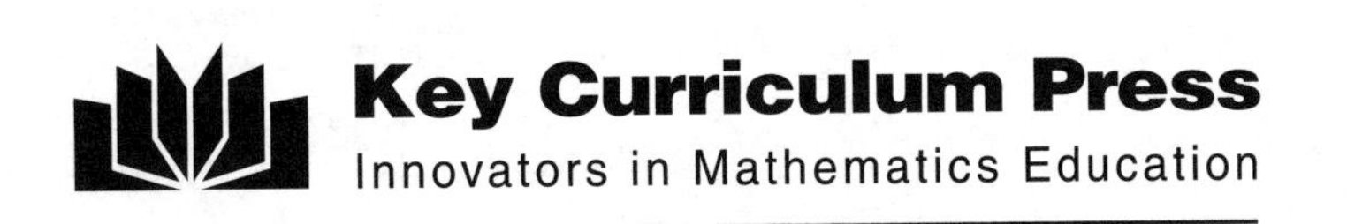

Comment Form

Please take a moment to provide us with feedback about this book. We are eager to read any comments or suggestions you may have. Once you've filled out this form, simply fold it along the dotted lines and drop it in the mail. We'll pay the postage. Thank you!

Your Name ____________________

School ____________________

School Address ____________________

City/State/Zip ____________________

Phone ____________________

Book Title ____________________

Please list any comments you have about this book.

____

Do you have any suggestions for improving the student or teacher material?

____

To request a catalog, or place an order, call us toll free at 800-995-MATH, or send a fax to 800-541-2242.
For more information, visit Key's website at www.keypress.com.

Please detach page, fold on lines and tape edge.

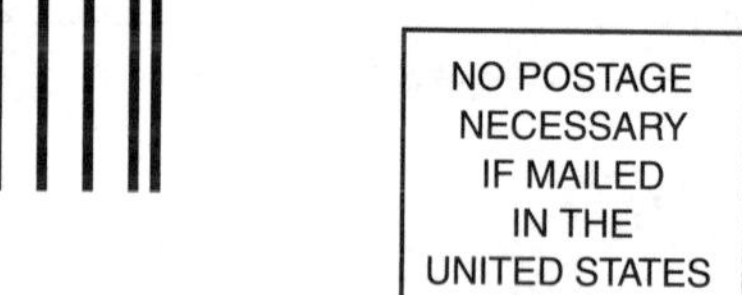

BUSINESS REPLY MAIL
FIRST CLASS PERMIT NO. 338 OAKLAND, CA

POSTAGE WILL BE PAID BY ADDRESSEE

KEY CURRICULUM PRESS
1150 65TH STREET
EMERYVILLE CA 94608-9740
ATTN: EDITORIAL

Notes

Notes